Basic Numerical Methods

An Introduction to Numerical Mathematics on a Microcomputer

R E Scraton

Reader in Numerical Analysis, University of Bradford

Edward Arnold

First published in Great Britain 1984 by
Edward Arnold (Publishers) Ltd, 41 Bedford Square, London WC1B 3DQ

Edward Arnold, 3 East Read Street, Baltimore, Maryland 21202, USA

Edward Arnold (Australia) Pty Ltd, 80 Waverley Road, Caulfield East,
Victoria 3145, Australia

Reprinted 1986

Scraton, R. E.
BASIC numerical methods.
1. Numerical analysis—Data processing
2. Basic (Computer program language)
I. Title
519.4 QA297

ISBN 0-7131-3521-2

Computer typeset by SB Datagraphics,
printed and bound by J. W. Arrowsmith Ltd, Bristol

Preface

When computers were first introduced, they were used almost entirely for the numerical solution of mathematical problems. The earliest computers were mostly to be found in the Mathematics departments of large universities. Nowadays computers, especially microcomputers, are big business. They are sold in quite large numbers to businessmen for accounting, data handling and word-processing; and in much larger numbers to children (of all ages) for games playing. Mathematicians now form an insignificant part of the total market, but they should be the last to complain about this: if computers were used only by mathematicians they would be many times their present price! For less than £200 one can now purchase machines which computer experts might haughtily dismiss as 'toys', but which are capable of serious and complex mathematical work: the author regularly uses a Sinclair Spectrum for his own research work.

One effect of the microcomputer revolution is that many budding mathematicians in their teens now have access to computing power which professional mathematicians could only dream about a few years ago. It is to these budding mathematicians, as well as budding scientists and engineers with a mathematical inclination, that this present book is addressed. Much of the book should be intelligible to the sixth-former studying A-level Mathematics, although as a whole the book is more suitable for the first year undergraduates. The author has tried to achieve a balance between computational practice and the underlying mathematical theory, but no doubt everyone has their own view on what the correct balance should be. The reader should not be afraid to skip some of the heavier mathematical work at a first reading; but must also remember that real competence in computational work cannot be attained without an understanding of the underlying theory.

It is assumed that the reader is sufficiently familiar with BASIC to understand the programs given throughout the book. These programs should be entered on the computer, tested (on as many problems as possible), and stored for future use. In these programs, REM statements have been used sparingly. It is rather tedious to type in lengthy and non-essential program lines, and in the author's experience many students simply leave them out. The most useful REM statements are those which a user inserts, which he or she feels are necessary for an understanding of the program: readers are invited to use these as liberally as they wish.

Although we have not specifically used the term in the text, we have tried to emphasise the importance of robustness in numerical methods. It is no use writing a program which works only for favourable problems. We should try to anticipate what might go wrong in less favourable circumstances. Ideally, we should like our programs to work for all relevant problems, but in numerical work we only rarely achieve this ideal; we have to be satisfied with programs which work 'nearly always'. No doubt some readers will take delight in finding problems for which the programs fail. If they go on to ask *why* the programs did not work—was the problem ill-posed, or was the program inadequate for the job—they will learn a great deal from this exercise.

A few exercises are given at the end of each chapter. Generally these are intended to lead the reader to further understanding of the subject, rather than merely provide repetitive examples. Answers to numerical questions are given at the end of the book, as well as illustrative programs for some of the programming exercises. Some of the exercises in the later chapters require fairly lengthy programs, and the reader who can cope with these successfully is well on the way to becoming a competent numerical programmer. The author will be delighted if just a few of his readers find this more satisfying than playing space invaders!

R E SCRATON
1984

Contents

Note on BASIC

Since nearly all microcomputers work in BASIC, it might be thought that a program written for one computer could readily be transferred to another. Alas this is not so! Every manufacturer has adapted and extended BASIC for his particular machine. Many of the extensions are to handle matters such as colour, sound or high resolution graphics, which were not envisaged by the original authors of BASIC; others incorporate programming structures which are advantageous although not strictly necessary. Whatever the cause, most programs require substantial rewriting before they can be used on a different machine.

BASIC is also intended to be an easily understood language. It uses everyday words such as 'IF', 'LET' and 'PRINT', and it should be possible to read through a program like a piece of English text. Unfortunately many modern programs make extensive use of instructions such as POKE, USR, DEFFIL or VDU, which are totally meaningless except to the users of a particular machine.

All of this creates problems in a book such as the present. The programs given here are intended to be both instructive and useful: they should show the reader how a particular numerical method is implemented, as well as enable him or her to use the method on whatever computer is available. In order to achieve this we have used a subset of BASIC which is readily understood and common to most modern versions of the language. The programs were written originally for the Commodore PET computer, but they will run with minimal alteration on most other computers. On some computers the programs can be shortened, for example by

(i) omitting LET wherever it occurs,
(ii) omitting either THEN or GOTO in a statement IF . . . THEN GOTO . . .,
(iii) omitting the brackets in a function such as SQR(X),
(iv) replacing NEXT R by just NEXT.

The reader will need to know which of these changes are permissible on his or her machine; if in doubt enter the program as written. No use has been made of structures such as IF . . . THEN . . . ELSE or REPEAT . . . UNTIL, as these are not available on all machines.

In this book we have used arrays with suffices starting at zero. Some computers—notably the Sinclair Spectrum—use arrays starting with suffix 1, so some amendment is needed to programs using arrays before they can be run on these machines. The easiest way to change the programs is to increase all array suffices by one: thus DIM A(N) is replaced by DIM A(N + 1), A(R, S) is replaced by A(R + 1, S + 1), and A(0) is replaced by A(1). This should always work, though it sometimes leads to rather inelegant programs. The Sinclair Spectrum again allows only single letter array names, so some of the arrays in Chapter 8 will need to be renamed: for instance K1(), K2(), K3() could be replaced by J(), K(), L(). Multiple dimension statements are not permitted on some machines, including the Spectrum: so DIM A(N, N + 1), X(N) must be written as DIM A(N, N + 1): DIM X(N).

It should also be realised that different machines do their arithmetic in different ways, so the numerical results given here (obtained on the PET) will not always be reproduced on other computers. Usually the difference will only be in the last significant figures, but where, for instance, we have deliberately taken the small difference between two nearly equal numbers, there could be substantial changes.

It is to be hoped that most readers will take the programs listed here as models to illustrate the implementation of a numerical procedure, and will rewrite them as necessary for their own computer. In this way they can take advantage of any permitted programming structures, and also make use of graphics or any other features they feel appropriate.

1

Numbers, Errors and Accuracy

1.1 Computer Accuracy

In the few years since microcomputers were introduced, they have become very much part of our everyday life. They may be found everywhere, from the amusement arcade to the managing director's office. Computers are especially important to the mathematician, for they enable him or her to carry out long and complicated sequences of operations which would otherwise be impossible, and this facility is no longer restricted to the privileged few who have access to an expensive main-frame computer.

A long sequence of calculations gives many opportunities for errors. There is no point in embarking on a complex calculation unless we can be reasonably sure that our answers will be correct, and that they will be sufficiently accurate for our purpose. The handbook for your microcomputer probably claims that the machine works to an accuracy of eight or nine significant figures, which is certainly accurate enough for most purposes. But this does not mean that every answer printed out by the computer is correct to this accuracy. On the Commodore PET, for instance, the instruction

```
PRINT INT(5-SQR(9))
```

gives the answer 1 instead of 2. The instruction

```
PRINT SQR(999999999)-SQR(999999998)
```

gives the answer

```
1.37Ø9Ø683E-Ø5
```

or 0.000 013 709 068 3. It is very tempting to take this answer as correct to the accuracy given, but as we shall see later the correct answer is 0.000 015 811 388 3.

At this stage you might feel tempted to return your computer to the manufacturer with a rude letter, but even the most expensive computers are capable of similar mistakes. We have to learn to use the computer in such a way as to avoid errors of this kind, and this is by no means easy. In this first chapter we shall see how to avoid some of the more common pitfalls.

1.2 How Numbers are Stored on the Computer

Most early calculators (and some early computers) worked in **fixed point** notation. They might be set, for instance, to display two figures before the decimal point and six figures after the decimal point. With this setting any answer larger than 100 would give an overflow error, whilst an answer less than 0.01 would be shown to only four significant figures. Nowadays nearly all scientific calculators work in **floating point** notation, or more strictly in **floating point decimal.** In this notation the display

1.234 567 89 +07

has to be interpreted as

$$1.234\ 567\ 89 \times 10^7 = 12\ 345\ 678.9.$$

These calculators can handle numbers from 10^{-99} to 10^{99}, all with the same number of significant figures. From the mathematician's viewpoint, the advantages of floating point over fixed point notation are obvious.

Computers work in **floating point binary** notation, although they display their results in fixed or floating point decimal. Most readers probably know a little about binary notation, and this will suffice for our present purpose. On most microcomputers the basic unit of storage is the **byte**, which consists of eight binary digits. We can consider a byte as holding an integer between 0 and 255 inclusive. All data held in the computer—numbers, strings or programs—are stored as combinations of bytes.

When a real number x is stored in the computer, it is first converted to the form

$$x = \pm a \times 2^b,$$

where a is a number such that $\frac{1}{2} \leqslant a < 1$ (known as the **mantissa**) and b is a positive, negative or zero integer (known as the **exponent**). Every real number (except zero) can be expressed uniquely in this form. Five bytes are used to store the number x, the first byte holding b and the remaining four holding a.

The first byte actually stores $b + 128$, so that b may take any value between -128 and $+127$. The computer can thus store numbers between 2^{-128} and 2^{127}, or roughly 2.9×10^{-39} and 1.7×10^{38}. Note that this is a smaller range than most scientific calculators can handle. The computer will stop with an appropriate error message if a number exceeds the upper limit. A number below the lower limit is simply taken as zero, with no error message, and this can sometimes cause unexpected results.

The mantissa a is stored as four bytes a_1, a_2, a_3, a_4, where

$$a = \frac{a_1}{256} + \frac{a_2}{256^2} + \frac{a_3}{256^3} + \frac{a_4}{256^4}.$$

This is very similar to the way in which we might write a decimal number to four decimal places, but here we are working in the scale of 256 instead of in the scale of 10. Since a lies between $\frac{1}{2}$ and 1, the first binary place of a_1 will always be 1, and need not be stored; instead this digit is used to store the sign of the number.

We can see that a is stored with a maximum error of $\frac{1}{2} \times 256^{-4} \simeq 1.2 \times 10^{-10}$. It is therefore correct to more than nine but less than ten decimal places. It follows that numbers are stored in the computer to an accuracy of somewhere between nine and ten significant figures.

1.3 Some Consequences of the Way Numbers are Stored

Apart from the addition, subtraction and multiplication of simple integers, nearly all operations on the computer yield only approximate answers, and it is important to be aware of this. Thus when the PET computer is asked for SQR(9) it does not obtain the exact answer 3, but a number slightly greater than 3. We are not immediately aware of this, however, for the instruction

```
PRINT SQR(9)
```

gives the answer 3. In fact the value for SQR(9) when rounded to nine (decimal) significant figures is 3.000 000 00, which the computer prints simply as 3. On the other hand 5 − SQR(9) gives a number slightly less than 2, so INT(5 − SQR(9)) gives 1 instead of 2. Other microcomputers may give the correct answer to this problem, but it is usually possible to find similar problems on which they fail.

It is important to realise that an instruction such as

```
100 IF SQR(9)=3 THEN PRINT "OK"
```

will not cause any printing; as far as the computer is concerned, SQR(9) is *not* equal to 3. In general we should avoid instructions beginning IF X = Y unless X and Y are simple integers: even if PRINT X and PRINT Y yield the same result the computer may not accept that X and Y are equal! It is better to write

```
100 IF ABS(X-Y)<Q THEN ...
```

where, say, Q = 1 E − 9 (i.e. $Q = 10^{-9}$).

Another unexpected result is illustrated by the following simple program:

```
10 REM SINE TABLE 1
20 FOR X=0 TO 1 STEP .1
30 PRINT X,SIN(X)
40 NEXT X
```

This is intended to produce a table of x and $\sin x$ from $x = 0$ to $x = 1$ in steps of 0.1 (or, as we usually write, for $x = 0(0.1)1$). In fact the output from this program looks like this:

```
0               0
.1              .0998334167
.2              .198669331
.3              .295520207
.4              .389418342
.5              .479425539
.6              .564642473
.7              .644217687
.8              .717356091
.9              .78332691
```

You will see that the output stops short at the value .9. After each circuit or the FOR–NEXT loop, the value of X is increased by 0.1, and the computer returns for another circuit of the loop provided that this new value of X is less than or equal to 1. The number 0.1 cannot be stored exactly on the computer, and in fact it is stored as a number slightly greater than 0.1; the effect of this is to bring the above program to a premature end. It is usually safer to make sure that the step-lengths in FOR–NEXT loops are integers, thus:

```
10 REM SINE TABLE 2
20 FOR R=0 TO 10
30 LET X=R/10
40 PRINT X,SIN(X)
50 NEXT R
```

This program gives the required result.

The next program is designed to calculate the binomial coefficient

$$\binom{n}{r} = \frac{n!}{r!(n-r)!}. \tag{1.1}$$

The subroutine starting in line 1ØØ calculates the value of M! and returns it as F; the rest of the program should be reasonably obvious.

```
1Ø REM BINOMIAL COEFFICIENT 1
2Ø INPUT"N,R";N,R
3Ø LET M=N:GOSUB 1ØØ:LET B=F
4Ø LET M=R:GOSUB 1ØØ:LET B=B/F
5Ø LET M=N-R:GOSUB 1ØØ:LET B=B/F
6Ø PRINT "BINOMIAL COEFFICIENT IS";B
7Ø STOP
1ØØ LET F=1:IF M<2 THEN RETURN
11Ø FOR S=2 TO M:LET F=F*S:NEXT S
12Ø RETURN
```

The program may be tested with small values of n and r, thus:

```
N,R? 4,2
BINOMIAL COEFFICIENT IS 6



N,R? 1Ø,3
BINOMIAL COEFFICIENT IS 12Ø



N,R? 2Ø,7
BINOMIAL COEFFICIENT IS 7752Ø
```

On the other hand, if we set $n = 35$, $r = 3$, we get the result

```
N,R? 35,3

?OVERFLOW ERROR IN 11Ø
```

even though the value of the required binomial coefficient is only 6545. The reason, of course, is that in the above program we calculate $n!$ as an intermediate result, and $35! \simeq 1.033 \times 10^{40}$, which is too large for the computer to handle. We can, however, write

$$\binom{n}{r} = \left(\frac{n}{r}\right)\left(\frac{n-1}{r-1}\right)\left(\frac{n-2}{r-2}\right)\cdots\left(\frac{n-r+1}{1}\right) \tag{1.2}$$

and using this formulation we get the program:

```
1Ø REM BINOMIAL COEFFICIENT 2
2Ø INPUT"N,R";N,R
3Ø LET B=1
4Ø IF R=Ø THEN GOTO 8Ø
5Ø FOR S=Ø TO R-1
6Ø LET B=B*(N-S)/(R-S)
7Ø NEXT S
8Ø PRINT "BINOMIAL COEFFICIENT IS";B
```

This gives the required result:

```
N,R? 35,3
BINOMIAL COEFFICIENT IS 6545
```

In fact this last program works for any values of n and r for which $\binom{n}{r}$ is within the capacity of the machine.

The formulae (1.1) and (1.2) are mathematically equivalent, but the latter proves to be more satisfactory from the computational viewpoint. It is often the case that the most obvious mathematical formulation is not the most satisfactory for computational work, and part of the art of numerical work is to choose the most suitable formulation.

1.4 Errors in Floating Point Arithmetic

We have already seen that a number x is stored as $\pm a \times 2^b$, where $\frac{1}{2} \leqslant a < 1$. Since a is stored to a length of four bytes (or equivalently to 32 binary places), there is a possible error in a in the interval $\pm\frac{1}{2} \times 256^{-4} = \pm 2^{-33}$. Thus the value of a stored in the computer is actually $a + E$, where $|E| \leqslant 2^{-33}$. It follows that the value of x stored in the computer is

$$\pm(a + E) \times 2^b = \pm a \times 2^b\left(1 + \frac{E}{a}\right) = x(1 + \varepsilon),$$

where

$$\varepsilon = \frac{E}{a}, \quad \text{and} \quad |\varepsilon| \leqslant \frac{1}{a} \times 2^{-33} \leqslant 2^{-32},$$

since $a \geqslant \frac{1}{2}$. So a number x is actually stored in the computer as $x(1 + \varepsilon)$, where $|\varepsilon| \leqslant \rho = 2^{-32}$. This result is true for any microcomputer which uses the five-byte representation of numbers described in Section 1.2. For other computers, or indeed for any calculator which uses floating point notation, the only difference is in the value of ρ: the smaller the value of ρ, the more accurate the machine. You might like to work out the value of ρ for your calculator.

Now suppose that we have two numbers x_1 and x_2, which are stored in the computer as $x_1(1 + \varepsilon_1)$ and $x_2(1 + \varepsilon_2)$, where $|\varepsilon_1| \leqslant \rho$ and $|\varepsilon_2| \leqslant \rho$. Suppose also that we wish to multiply these two numbers together to obtain a number $y = x_1x_2$. The value obtained for y will be

$$\begin{aligned} x_1(1 + \varepsilon_1)x_2(1 + \varepsilon_2) &= x_1x_2(1 + \varepsilon_1 + \varepsilon_2 + \varepsilon_1\varepsilon_2) \\ &\simeq y(1 + \varepsilon_1 + \varepsilon_2), \end{aligned}$$

since $\varepsilon_1\varepsilon_2$ is small compared with the other terms involved. Thus y is obtained as $y(1 + \delta)$, where $\delta \simeq \varepsilon_1 + \varepsilon_2$, and so $|\delta| \leqslant 2\rho$. This gives an upper limit to the magnitude of the error in y, but in many cases the error will be much smaller than this: for instance if ε_1 and ε_2 are of opposite signs they will partially cancel one another. Nevertheless we must allow for the possibility of a slight loss of accuracy whenever we multiply two numbers together, since the limit for $|\delta|$ is twice the limit for $|\varepsilon|$.

If, instead, we calculate $y = x_1/x_2$, the value obtained is

$$\frac{x_1(1+\varepsilon_1)}{x_2(1+\varepsilon_2)} = \frac{x_1}{x_2}(1+\varepsilon_1)(1-\varepsilon_2+\varepsilon_2^2-\varepsilon_2^3+\ldots),$$

using the binomial expansion of $(1+\varepsilon_2)^{-1}$. Ignoring terms involving products of two or more ε's, this gives

$$y(1+\varepsilon_1-\varepsilon_2).$$

Thus again y is obtained as $y(1+\delta)$, where in this case $\delta \simeq \varepsilon_1 - \varepsilon_2$. Since ε_1 and ε_2 can be either positive or negative, we again have $|\delta| \leqslant 2\rho$; thus the possible loss of accuracy on division is similar to that on multiplication.

Addition and subtraction are a little more difficult to handle. If $y = x_1 + x_2$ then y is computed as

$$x_1(1+\varepsilon_1) + x_2(1+\varepsilon_2) = x_1 + x_2 + x_1\varepsilon_1 + x_2\varepsilon_2 = y(1+\delta),$$

where

$$\delta = \frac{x_1\varepsilon_1 + x_2\varepsilon_2}{x_1 + x_2}. \tag{1.3}$$

We know that $-\rho \leqslant \varepsilon_1 \leqslant \rho$, $-\rho \leqslant \varepsilon_2 \leqslant \rho$; so if x_1 and x_2 are both positive we can write

$$-\frac{x_1\rho + x_2\rho}{x_1 + x_2} \leqslant \delta \leqslant \frac{x_1\rho + x_2\rho}{x_1 + x_2}$$

or $|\delta| \leqslant \rho$. Since the limit for $|\delta|$ is the same as the limit for $|\varepsilon|$ there is no loss of accuracy when two positive numbers are added together. We can easily extend the above argument to the case when x_1 and x_2 are both negative, but not to the case when x_1 and x_2 are of opposite signs, i.e. when we are either effectively doing a subtraction.

If x_2 is a number close to $-x_1$, the denominator $x_1 + x_2$ in equation (1.3) is nearly zero, so δ could be indefinitely large. It follows that we can get a substantial loss of accuracy when 'subtracting' two nearly equal numbers. It is probably true to say that this is the most common cause of loss of accuracy in computational work. 'Subtraction' in this sense may either be the addition of two numbers of opposite sign or the subtraction of two numbers of the same sign. Whenever we use the symbols + or − in a computer program we should remember that this could be a possible source of error!

In Section 1.1 we mentioned the evaluation of SQR(999999999) − SQR(999999998). This is a clear example of the subtraction of two nearly equal numbers. The instruction

```
PRINT SQR(999999999), SQR(999999998)
```

gives the answers

```
31622.7766          31622.7766
```

showing that the two numbers are identical to nine significant figures. Since the computer holds these numbers only to a little more than nine significant figures, it is obvious that we can expect no accuracy at all when we subtract them; but the computer nevertheless insists on giving us a full nine-figure answer. What may appear obvious when presented in this way will be far from obvious when it occurs in the middle of a lengthy computer program; yet it could well destroy any accuracy in our final results. It is imperative that we formulate our problems so as to avoid any possibility of subtracting nearly equal numbers. We can reformulate the above problem by setting $p^2 = 999\,999\,999$, $q^2 = 999\,999\,998$, then

$$p - q = \frac{p^2 - q^2}{p + q} = \frac{1}{p + q};$$

we can thus evaluate $p - q$ without any loss of accuracy by means of the instruction

```
PRINT 1/(SQR(999999999)+SQR(999999998))
```

which gives the correct answer

```
1.58113883E-Ø5
```

1.5 Simple Interval Arithmetic

So far we have been concerned only with errors which arise in the computer, but we must be equally aware of errors in the data which we feed into the computer. If we ask the computer to evaluate 4.56/1.23 we shall get the answer 3.707 317 07, but it would be futile to quote this answer to nine significant figures if the numbers 4.56 and 1.23 are accurate only to three. If this is the case 4.56 denotes a number between 4.555 and 4.565, whilst 1.23 denotes a number between 1.225 and 1.235. Their quotient could be as small as $4.555/1.235 = 3.6883$, or as large as $4.565/1.225 = 3.7265$. Thus the answer lies between 3.6883 and 3.7265, and is accurate to rather less than three significant figures.

What we have just done is a simple example of **interval arithmetic**. It is convenient for this purpose to introduce the notation $\{p, q\}$ to mean 'a number between p and q'. We can then write

$$4.56 = \{4.555, 4.565\} \qquad 1.23 = \{1.225, 1.235\}$$

and so

$$4.56/1.23 = \{3.6883, 3.7265\}$$

$$4.56 \times 1.23 = \{5.5799, 5.6378\}$$

$$4.56 + 1.23 = \{5.78, 5.80\}$$

$$4.56 - 1.23 = \{3.32, 3.34\}.$$

Check the above answers for yourself, and make sure you understand how they are worked out. Note that in multiplication and addition we obtain the lower limit by combining 4.555 with 1.225, whereas in division and subtraction we combine 4.555 with 1.235. In every case we have to decide which combination gives the smallest possible result, and which gives the largest.

The idea is easily extended to more complicated expressions, for example

$$\frac{1.23 + 4.56}{2.78 \times (9.87 - 8.72)}$$

$$= \frac{\{1.225, 1.235\} + \{4.555, 4.565\}}{\{2.775, 2.785\} \times (\{9.865, 9.875\} - \{8.715, 8.725\})}$$

$$= \frac{\{5.78, 5.80\}}{\{2.775, 2.785\} \times \{1.14, 1.16\}}$$

$$= \frac{\{5.78, 5.80\}}{\{3.1635, 3.2306\}}$$

$$= \{1.7891, 1.8334\}.$$

The answer, in this case, is accurate to only two significant figures, and we should quote it as 1.8.

It is often believed that data which are accurate to three significant figures will give answers accurate to three significant figures. Whilst this may be a rough-and-ready rule in some circumstances, it is certainly not strictly true, as can be seen in the above examples. The most obvious deviation from this rule is again in the subtraction of nearly equal numbers; for example

$$9.87 - 9.86 = \{9.865, 9.875\} - \{9.855, 9.865\}$$
$$= \{0.00, 0.02\},$$

giving virtually no accuracy in the answer. There are many other instances, for example

$$\tan 89.7° = \{\tan 89.65°, \tan 89.75°\} = \{163.7, 229.2\},$$

with an accuracy of only one significant figure.

Interval arithmetic gives us a foolproof way of assessing the possible error in data manipulation. It deals only with extreme values, and in practice most answers will lie well within the limits obtained. Clearly it is impractical to do all data manipulation using interval arithmetic, but used occasionally it can help to identify sources of error.

Exercises

1 Obtain both roots of the quadratic equation

$$x^2 - 23\,456x + 7 = 0$$

correct to nine significant figures.

2 Statisticians often need to calculate the binomial probability

$$P = \binom{n}{r} p^r (1-p)^{n-r}, \quad (0 \leqslant r \leqslant n, \quad 0 < p < 1).$$

Write a program to evaluate P for any specified values of n, r and p. Try to ensure that you get no overflow errors, and that you do not get the answer zero unless $P < 10^{-38}$.

Test your program with $n = 200$, $r = 100$, $p = 0.85$.
(Answer: 3.22×10^{-31}.)

3 Write a program to evaluate the sum of the power series

$$1 + \frac{x}{1!} + \frac{x^2}{2!} + \frac{x^3}{3!} + \ldots$$

for any real value of x. Arrange for the summation to stop as soon as a term is less than 10^{-9} in magnitude. Run your program with a variety of values of x, e.g. ± 0.1, ± 1, ± 10, ± 20, ± 50, and compare your answer with the correct value of e^x. Comment on your results.

4 Find, correct to nine significant figures,

(i) $999\,999\,999^{\frac{1}{3}} - 999\,999\,998^{\frac{1}{3}}$,

(ii) $999\,999\,999^{\frac{1}{4}} - 999\,999\,998^{\frac{1}{4}}$.

5 In the following equations, all numbers have been rounded to three significant figures. How accurately can x be determined in each case?

(i) $\cos x = 0.999$

(ii) $\cosh x = 1.01$

(iii) $\tanh x = 0.999$

(iv) $x = \tan 1.56$

(v) $x = 0.250 \times 0.125 - 0.200 \times 0.167$

(Note: throughout this book trigonometric functions are to be taken in radians, unless specifically stated otherwise.)

6 In the simultaneous equations

$$7x - 2y = 6$$
$$x + 0.28y = 2$$

the value 0.28 is correct to two decimal places, but all other numbers are exact. How accurately can x and y be determined? Repeat for the equations

$$7x + 2y = 6$$
$$x + 0.28y = 2.$$

7 You have borrowed £1000 from a bank which charges compound interest at the rate of 2% per month. Using four-figure logarithm tables, work out how much you owe the bank after ten years. How accurate is your answer?

2

Iterative Methods

2.1 Introduction

As an introduction to the idea of iterative methods, here is a simple party trick to play on your calculator. Set your calculator to work in the 'degrees' mode and enter any number you like. Press the COS button four times. Your answer is now 0.999 847 741. If you experiment with this trick, you will find that you get exactly the same number at the end no matter what number you start with. You will also find that if you press the COS button more than four times, the answer 0.999 847 741 will be displayed repeatedly. So what is magic about this number?

If we let $\alpha = 0.999\ 847\ 741$, then we have just shown that $\cos \alpha°$ is also equal to α, at least to the accuracy of the calculator. In other words α is a root of the equation

$$\cos x° = x. \tag{2.1}$$

This in itself is interesting. If you had been asked to solve the equation (2.1) you probably would not know where to begin, except possibly by some means of trial and error; yet almost by accident we have solved the equation to the full accuracy of our calculator. We often want to solve equations like (2.1)—although usually a good deal more complicated—so it is worth analysing the procedure we have just used to see if it is of more general application.

We started by setting any number x_0 on the calculator; say for instance $x_0 = -22.2222$. We then calculate in turn

$$\begin{aligned} x_1 &= \cos x_0° = 0.925\ 724\ 116 \\ x_2 &= \cos x_1° = 0.999\ 869\ 479 \\ x_3 &= \cos x_2° = 0.999\ 847\ 735 \\ x_4 &= \cos x_3° = 0.999\ 847\ 741 \\ x_5 &= \cos x_4° = 0.999\ 847\ 741, \text{ etc.} \end{aligned}$$

This is an example of an **iterative process**. Starting with any appropriate value for x_0 we set

$$x_{r+1} = F(x_r), \quad r = 0, 1, 2, 3, \ldots, \tag{2.2}$$

for some function F. We continue until we reach a stage where x_{r+1} is indistinguishable from x_r to the accuracy to which we are working. This value is then a root of the equation

$$x = F(x). \tag{2.3}$$

2.2 Solution of a Simple Equation

We will now try another example, this time making use of the computer. We define the iterative process

$$x_{r+1} = \tfrac{1}{2}x_r^2 + \tfrac{1}{4}, \qquad (2.4)$$

and implement it by means of the following program:

```
10 REM ITERATIVE PROCESS 1
20 INPUT"X0";X
30 LET R=0
40 LET R=R+1
50 LET X=.5*X*X+.25
60 PRINT R,X
70 GOTO 40
```

If we set $x_0 = 1$, the output looks like this:

```
X0? 1
 1          .75
 2          .53125
 3          .391113281
 4          .326484799
 5          .303296162
 6          .295994281
 7          .293806307
 8          .293161073
 9          .292971707
 10         .292916211
 11         .292899953
 12         .292895191
 13         .292893797
 14         .292893388
 15         .292893268
 16         .292893233
 17         .292893223
 18         .29289322
 19         .292893219
 20         .292893219
 ...................
 ...................
```

Note that in writing the above program it is not necessary to use an array X(R) to store the value of x_r. Once we have computed x_{r+1} we have finished with x_r, so both can be stored as X. It is important to remember points like this, or we could soon fill the computer's memory with unnecessary arrays.

One defect in the above program is that it runs indefinitely until we press the STOP or BREAK key. A better program would be:

```
10 REM ITERATIVE PROCESS 2
20 INPUT"X0";X
30 LET R=0:LET Q=1E-9
40 LET R=R+1
50 LET Y=X
60 LET X=.5*X*X+.25
70 PRINT R,X
80 IF ABS(Y-X)>Q THEN GOTO 40
90 PRINT:PRINT "ROOT IS";X
```

In this version, the variable Y is used to store the value of x_r before X is updated to x_{r+1} in line 60. Line 80 ensures that the program stops when X and Y differ by less than 10^{-9} (note that we do not simply test whether X = Y, for the reasons explained in the last chapter). When run with $x_0 = 1$, the program stops when R = 19, thus:

```
...................
...................

16          .292893233
17          .292893223
18          .29289322
19          .292893219

ROOT IS .292893219
```

The number just obtained is a root of the equation

$$x = \tfrac{1}{2}x^2 + \tfrac{1}{4},$$

which may be rewritten as the quadratic equation

$$2x^2 - 4x + 1 = 0.$$

The roots of this equation are $1 \pm \sqrt{0.5} = 0.292\,893\,219,\ 1.707\,106\,781$. You may be wondering why we obtained one root of the quadratic equation rather than the other, and you may imagine that this has something to do with the choice of starting value x_0. We should therefore run the program with some different starting values; but if we set the starting value equal to 2 we get a nasty surprise:

```
XØ? 2
 1          2.25
 2          2.78125
 3          4.11767578
 4          8.72762692
 5          38.3357358
 6          735.Ø64321
 7          27Ø16Ø.Ø28
 8          3.649322Ø3E+1Ø
 9          6.65877565E+2Ø

?OVERFLOW ERROR IN 6Ø
```

In this case the process has diverged! We do not wish to delve too deeply into Pure Mathematics here, but the reader is probably aware that a sequence of numbers $\{x_r\}$, however defined, may either converge to a limit or diverge as $r \to \infty$. The same is true of a sequence produced by an iterative process, but clearly an iterative process has no computational value unless it converges.

2.3 Graphical Representation

You should now run the last program with many different values of x_0 and see what happens. You will find that the process converges if $|x_0| < 1.707\,106\,78$ (the larger root of the quadratic equation), but convergence is always to the smaller root 0.292 893 219. On the other hand if $|x_0| > 1.707\,106\,78$, the process diverges. This convergence and divergence is illustrated graphically in Figs 2.1 and 2.2. In these diagrams, we show the graphs of $y = x$ and $y = \tfrac{1}{2}x^2 + \tfrac{1}{4}$. The graphs intersect at A and B, corresponding to the two roots of the equation. Given the value of x_0, the value of x_1 is the height of the point P_1, where the vertical through x_0 cuts the curve $y = \tfrac{1}{2}x^2 + \tfrac{1}{4}$. The point Q_1 on the line $y = x$ has coordinates (x_1, x_1), and the height of P_2 is the value of x_2. As the process proceeds we move along the 'staircase' $P_1Q_1P_2Q_2P_3Q_3$.... In Fig. 2.1, we get closer and closer to the point A, showing that the process converges to the smaller root. In Fig. 2.2, the staircase quickly moves off the paper, showing a rapid divergence. Unless we start exactly at the point B, there is no way of getting convergence to the larger root.

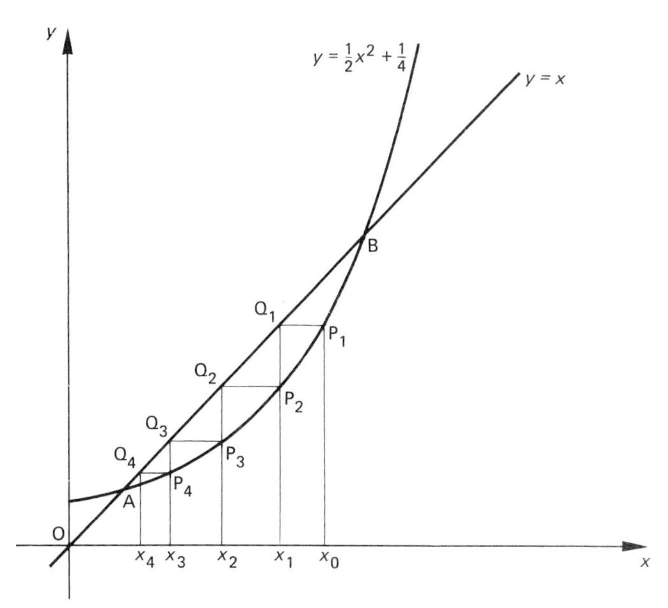

Fig. 2.1 Convergence of the iterative process (2.4)

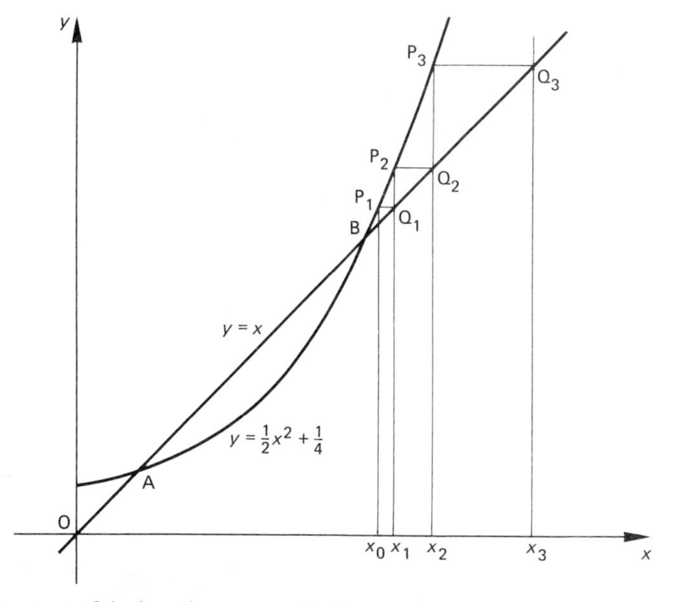

Fig. 2.2 Divergence of the iterative process (2.4)

2.4 Some More Iterative Processes

In the above work we have been trying to solve the quadratic equation

$$2x^2 - 4x + 1 = 0. \tag{2.5}$$

The iterative process (2.4) was obtained by rewriting this equation in the form

$$x = \tfrac{1}{2}x^2 + \tfrac{1}{4}.$$

In order to obtain an iterative process to solve a particular equation we have to rewrite the equation in the form $x = F(x)$, but there are clearly many ways in which we can do this. We could, for instance, write equation (2.5) in the form

$$x = \sqrt{(2x - \tfrac{1}{2})},$$

giving the iterative process

$$x_{r+1} = \sqrt{(2x_r - \tfrac{1}{2})}.$$

In order to implement this process, we merely have to change line 6Ø in the last program to

```
6Ø LET X=SQR(2*X-.5)
```

This should convince you of the beautiful simplicity of iterative processes from the point of view of the programmer!

Try running this amended program with various values of x_0. You will find that for any value of x_0 greater than 0.292 893 219 the process converges to the larger root 1.707 106 78. For values of x_0 less than 0.292 893 219 we get, sooner or later, the error message

```
?ILLEGAL QUANTITY ERROR IN 6Ø
```

due to trying to take the square root of a negative number. Try to illustrate this behaviour graphically in the manner of Figs 2.1 and 2.2.

At this stage you should try to think up some iterative processes for yourself, and investigate their convergence/divergence for various starting values. All you have to do is change line 6Ø in the last program, and leave the computer to do the donkey work. For illustration we shall consider two more equations here. First, consider another quadratic equation

$$x^2 + x - 6 = 0, \tag{2.6}$$

whose roots are known to be 2 and -3. Our previous approach suggests the iterative process

$$x_{r+1} = 6 - x_r^2,$$

but if you try this you will find that it diverges for all values of x_0 except $x_0 = 2$ and $x_0 = -3$. Another way of rewriting equation (2.6) suggests the iterative process

$$x_{r+1} = \sqrt{(6 - x_r)}.$$

This will be found to converge to the root 2 for any $x_0 \leqslant 6$; for $x_0 > 6$ we get an immediate ILLEGAL QUANTITY ERROR.

Secondly, consider the equation

$$x + \ln x = 0 \tag{2.7}$$

An obvious iterative process is

$$x_{r+1} = -\ln x_r,$$

which we implement by changing line 6Ø to

```
6Ø LET X=-LOG(X)
```

Whatever starting value we use, we eventually get

```
?ILLEGAL QUANTITY ERROR IN 6Ø
```

caused by trying to take the logarithm of a negative number. We can also write equation (2.7) as

$$x = e^{-x}$$

to give the iterative process

$$x_{r+1} = e^{-x_r}.$$

This converges to the root 0.567 143 290 for most values of x_0, but we get an overflow error if x_0 is taken as a large negative number.

It should be clear by this stage that there are two major difficulties in the implementation of an iterative process:

(i) the process may not converge;
(ii) even if it does converge, it may converge to the wrong root.

In the next chapter we shall describe some standard iterative processes which will usually give the required root of any equation; but first we need to do some theoretical work.

2.5 Theoretical Investigation of Convergence

Suppose we have an iterative process

$$x_{r+1} = F(x_r)$$

and we want to know whether or not this process converges to a given root α of the equation

$$x = F(x).$$

We will suppose for the moment that x_r is fairly close to α, so that we can write

$$x_r = \alpha + \varepsilon_r$$

where ε_r is small. If ε_r gets smaller still as r increases, then the process converges to α. If it gets larger then the process diverges.

We have

$$\begin{aligned} x_{r+1} &= F(x_r) \\ &= F(\alpha + \varepsilon_r) \\ &= F(\alpha) + \varepsilon_r F'(\alpha) + \tfrac{1}{2}\varepsilon_r^2 F''(\alpha) + \ldots \end{aligned} \tag{2.8}$$

by using a Taylor series. Now $\alpha = F(\alpha)$, since α is a root of the equation; and if ε_r is small we may neglect ε_r^2, to give

$$x_{r+1} \simeq \alpha + \varepsilon_r F'(\alpha)$$

or

$$\varepsilon_{r+1} \simeq \varepsilon_r F'(\alpha).$$

It is immediately obvious that ε_r gets smaller as r increases if and only if $|F'(\alpha)| < 1$, so this is the condition for convergence. But note that we have assumed in the above work that ε_r is small. Thus the process will converge if $|F'(\alpha)| < 1$ *provided that we start sufficiently close to the required root.* The process may converge even if we do not start close to the required root, but we cannot deduce this from the above treatment.

The result just obtained is not immediately useful, because we need to evaluate the function F' at the root α; and of course we do not know the value of α before we start! Nevertheless we often have a rough idea of the value of the root, and this will generally suffice. For example, the iterative process

$$x_{r+1} = \tfrac{1}{2}x_r^2 + \tfrac{1}{4}$$

gives $F'(x) = x$. Even if we know only that one root lies between 0 and 1 and another between 1 and 2, we can immediately say that the process will converge to the former but not to the latter. For the process

$$x_{r+1} = \sqrt{(2x_r - \tfrac{1}{2})},$$

we have

$$F'(x) = (2x - \tfrac{1}{2})^{-\frac{1}{2}}.$$

Thus $F'(0.3) \simeq 3.2$ and $F'(1.7) \simeq 0.6$; the process will converge to a root near 1.7, but not to a root near 0.3.

2.6 The Rate of Convergence

You will already have noticed that some iterative processes converge more rapidly than others, and naturally we prefer those which converge rapidly. Some processes converge very slowly indeed. Consider, for instance, the process

$$x_{r+1} = -\frac{1}{5}\left(7x_r - 24 + \frac{6}{x_r}\right). \tag{2.9}$$

If this converges, its limit satisfies

$$-5x = 7x - 24 + \frac{6}{x}$$

which simplifies to our old friend

$$2x^2 - 4x + 1 = 0.$$

Test this process by changing line 6Ø to

```
6Ø LET X=-(7*X-24+6/X)/5
```

and using the starting value $x_0 = 2$. You will find that the program never stops but eventually (after about 1300 steps) prints out alternately the values 1.707 106 76 and 1.707 106 81. Oscillations of this kind quite often occur with slowly convergent processes; the present process is not sufficiently sensitive to give more than eight-figure accuracy.

For the process (2.9),

$$F'(x) = -\frac{1}{5}\left(7 - \frac{6}{x^2}\right)$$

and so

$$F'(1.707\ 106\ 78) = -0.988\ 225\ 099.$$

The process converges since $|F'(\alpha)| < 1$, but because $|F'(\alpha)|$ is so close to 1 the convergence is very slow; ε_{r+1} is smaller in magnitude than ε_r, but only just!

Now consider the process

$$x_{r+1} = \frac{2x_r^2 - 1}{4(x_r - 1)}. \tag{2.10}$$

A little algebraic manipulation will show that this also solves the quadratic equation (2.5). Try running this process, again with $x_0 = 2$, and you will find that you get the root 1.707 106 78 after only five steps. The contrast between this process and the last could hardly be more marked. In the present case

$$F(x) = \frac{2x^2 - 1}{4(x - 1)}$$

and so

$$F'(x) = \frac{(x-1)(4x) - (2x^2 - 1)(1)}{4(x-1)^2}$$

$$= \frac{2x^2 - 4x + 1}{4(x-1)^2}.$$

For either root of the quadratic equation, the numerator of $F'(x)$ is zero. Thus, if α is either root, $|F'(\alpha)|$ is as small as it possibly can be, and we can expect very rapid convergence.

If we look back to equation (2.8), we shall see that, if $F'(\alpha) = 0$, then

$$x_{r+1} = F(\alpha) + \tfrac{1}{2}\varepsilon_r^2 F''(\alpha) + \ldots$$

and so

$$\varepsilon_{r+1} \simeq \tfrac{1}{2}\varepsilon_r^2 F''(\alpha).$$

In this case ε_{r+1} is a multiple of the *square* of ε_r; so if ε_r is small then ε_{r+1} is much smaller—in fact the number of decimal places accurate will roughly double at each stage of the iteration.

If an iterative process is such that

$$\varepsilon_{r+1} \simeq A\,\varepsilon_r^n,$$

where A is a constant, the process is said to be nth-order. Most of the processes we have met so far have been first-order, but process (2.10) is second-order. More specifically,

(i) a process is first-order if $F'(\alpha) \neq 0$ (but converges only if $|F'(\alpha)| < 1$);
(ii) a process is second-order if $F'(\alpha) = 0$ but $F''(\alpha) \neq 0$;
(iii) a process is third-order if $F'(\alpha) = 0$ and $F''(\alpha) = 0$, but $F'''(\alpha) \neq 0$;

and so on. Note that processes of order higher than the first always converge if we start sufficiently close to the required root. Most iterative processes in common use are either first-order or second-order; our experience above shows that it is worth taking a little trouble to find a second-order process.

2.7 Acceleration of Convergence

Sometimes we spend a long time developing a convergent iterative process only to find that, like process (2.9), it is too slowly convergent to be of practical use. All is not lost, however, for we can usually speed up the convergence of any first-order method.

We showed in Section 2.5 that for any first-order process we can write

$$\varepsilon_{r+1} \simeq \lambda\varepsilon_r,$$

where for brevity we have taken $\lambda = F'(\alpha)$. It follows that

$$\varepsilon_{r+2} \simeq \lambda\varepsilon_{r+1} \simeq \lambda^2\varepsilon_r,$$

and so

$$x_r = \alpha + \varepsilon_r$$

$$x_{r+1} \simeq \alpha + \lambda\varepsilon_r$$

$$x_{r+2} \simeq \alpha + \lambda^2\varepsilon_r.$$

Subtracting we get

$$x_{r+1} - x_r \simeq (\lambda - 1)\varepsilon_r$$

$$x_{r+2} - x_{r+1} \simeq \lambda(\lambda - 1)\varepsilon_r$$

and

$$x_{r+2} - 2x_{r+1} + x_r \simeq (\lambda - 1)^2\,\varepsilon_r.$$

Thus

$$\frac{(x_{r+2} - x_{r+1})^2}{x_{r+2} - 2x_{r+1} + x_r} \simeq \lambda^2\varepsilon_r$$

and hence

$$\alpha \simeq x_{r+2} - \frac{(x_{r+2} - x_{r+1})^2}{x_{r+2} - 2x_{r+1} + x_r}. \tag{2.11}$$

In the above we have neglected terms of order ε_r^2, and the value of α given by equation (2.11) has an error of this order. By making use of this result we can effectively convert any first-order process into a second-order process.

The procedure used to accelerate the convergence of a first-order process is as follows:

(i) Start with a suitable x_0.
(ii) Take $x_1 = F(x_0)$.
(iii) Take $x_2 = F(x_1)$.
(iv) Take $x_3 = x_2 - \dfrac{(x_2 - x_1)^2}{x_2 - 2x_1 + x_0}$.
(v) Take $x_4 = F(x_3)$.
(vi) Take $x_5 = F(x_4)$.
(vii) Take $x_6 = x_5 - \dfrac{(x_5 - x_4)^2}{x_5 - 2x_4 + x_3}$, and so on.

This procedure is implemented in the program below. This is designed to accelerate the convergence of process (2.9), but it can easily be adapted for any other process by changing the function definition in line 2Ø. Note that the variable A is used to store the values of $x_0, x_3, x_6, \ldots$; B to store the values of $x_1, x_4, x_7, \ldots$; and C to store the values of $x_2, x_5, x_8, \ldots$.

```
1Ø REM ACCELERATION
2Ø DEF FNF(X)=-(7*X-24+6/X)/5
3Ø INPUT"XØ";A
4Ø LET R=Ø
5Ø PRINT:PRINT R,A
6Ø LET B=FNF(A):PRINT R+1,B
7Ø LET C=FNF(B):PRINT R+2,C
8Ø LET A=C-(C-B)*(C-B)/(C-2*B+A)
9Ø LET R=R+3:GOTO 5Ø
```

Starting with $x_0 = 2$, this program gives the output:

```
XØ? 2

Ø          2
1          1.4
2          1.98285714

3          1.69565218
4          1.71839465
5          1.6959213

6          1.7Ø7Ø91Ø9
7          1.7Ø712229
8          1.7Ø7Ø9146

9          1.7Ø71Ø678
1Ø         1.7Ø71Ø678
11         1.7Ø71Ø678

?DIVISION BY ZERO ERROR IN 8Ø
```

As will be seen, the program stops with a DIVISION BY ZERO ERROR. At this stage, A, B and C are all equal and so C – 2*B + A is zero. You might like to amend the program to prevent division by zero, and so as to stop the program when sufficient accuracy has been obtained. But quite clearly the acceleration procedure is a substantial improvement on the original version of process (2.9), although still not as fast as the second-order process (2.10).

Exercises

1 Try the following iterative processes on the computer with various values of x_0. Comment on the convergence and the limit obtained. In at least one case, illustrate the convergence/divergence graphically.

(i) $x_{r+1} = \cos x_r$;
(ii) $x_{r+1} = \frac{1}{2}(x_r^3 + 1)$;
(iii) $x_{r+1} = x_r^3 + 1$;
(iv) $x_{r+1} = -(1 - x_r)^{\frac{1}{3}}$.

2 (a) The pair of variables (x_r, y_r) is produced by the process

$$x_{r+1} = \tfrac{1}{3}(1 - 2y_r)$$
$$y_{r+1} = -\tfrac{1}{3}(16 + x_r).$$

Write a program to tabulate (x_r, y_r) as r increases, starting with arbitrary values for (x_0, y_0). Show that the values of (x_r, y_r) converge to limiting values (α, β). What do α and β represent?

(b) Let $x_r = \alpha + \delta_r$, $y_r = \beta + \varepsilon_r$ in the above process, and find expressions for x_{r+2}, y_{r+2}. Hence prove theoretically that the process converges.

3 (a) Sketch a graph of the function f, where

$$f(x) = 2x^3 + 3x^2 + 4x + 5$$

and hence show that the equation $f(x) = 0$ has only one real root, lying between -1.3 and -1.4.

(b) Show that the iterative process

$$x_{r+1} = \frac{1}{31}\left(24x_r^2 + 67x_r + 48 + \frac{60}{x_r}\right),$$

if it converges, gives a root of the above equation. Try this process with various starting values x_0. Show that convergence is obtained only if x_0 is fairly close to the required root, and even then convergence is slow.

(c) Investigate the convergence of the process in (b) theoretically, and explain why convergence is slow.

(d) Use the iterative process in (b) in conjunction with the acceleration procedure to obtain the required root as accurately as you can.

3
Solution of Equations

3.1 The General Problem

In this chapter we shall consider some standard methods for solving an equation

$$f(x) = 0. \tag{3.1}$$

We have already seen in the last chapter that such an equation can be rewritten in the form

$$x = F(x)$$

in a variety of ways (be careful not to confuse the functions f and F) and then treated by an iterative method. The methods dealt with in this chapter will all be iterative in nature, but we shall be putting the topic on a more formal basis.

An equation such as (3.1) may have no real roots, one real root, many real roots, or even an infinite number of real roots. We may want to find all the real roots, some of them (say all the positive roots) or just one particular root. The equation may also have complex roots, and in some circumstances we may want to find these, but in the present volume we shall be content with real roots.

Before we start on any serious computational work, it is a good idea to sketch a graph of $f(x)$ and see where this graph cuts the x-axis. This tells us how many real roots there are and gives us an approximate idea of their values. Thus if the graph of $f(x)$ looked like Fig. 3.1 we should see that there are three real roots, in the intervals (1,2), (3,4), (5,6). Usually quite a rough sketch will do for this purpose.

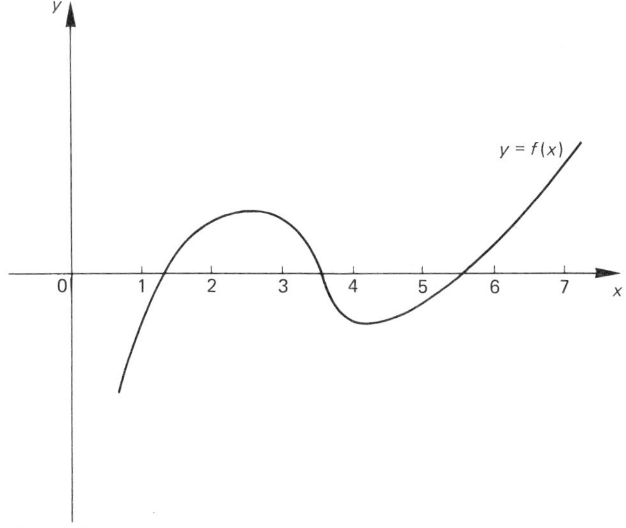

Fig. 3.1 Approximate determination of the roots of an equation

3.2 The Interval Bisection Method

The interval bisection method is one of the most obvious ways of finding a root. Suppose we know that $f(x) = 0$ has one root between $x = a$ and $x = b$; then $f(a)$ and $f(b)$ have opposite signs (assuming of course that the graph of $f(x)$ is continuous between a and b; we must at least make that assumption!) Now let c be halfway between a and b, i.e. $c = \frac{1}{2}(a + b)$, and evaluate $f(c)$. If $f(c)$ has the same sign as $f(a)$, then the root lies between c and b; otherwise the root lies between a and c. In either case we have reduced the interval in which the root lies to one-half its original width. We repeat the process, reducing the interval to $\frac{1}{4}$, $\frac{1}{8}$, $\frac{1}{16}$, ... of its original width, until we have obtained the root to whatever accuracy we require.

The program below uses this method to solve the equation

$$e^{-x} - x = 0, \tag{3.2}$$

but by changing line 2Ø it can be used for any equation. In this program $f(a)$ and $f(c)$ are denoted by F and G respectively. We test whether two numbers are of opposite sign by testing whether their product is negative.

```
1Ø REM INTERVAL BISECTION METHOD
2Ø DEF FNF(X)=EXP(-X)-X
3Ø LET Q=1E-9
4Ø INPUT"INITIAL LIMITS";A,B
5Ø LET F=FNF(A)
6Ø IF F*FNF(B)<Ø THEN GOTO 9Ø
7Ø PRINT "F(A) AND F(B) HAVE SAME SIGN: TRY AGAIN"
8Ø GOTO 4Ø
9Ø LET C=(A+B)/2:LET G=FNF(C)
1ØØ IF F*G<Ø THEN LET B=C:GOTO 12Ø
11Ø LET A=C:LET F=G
12Ø IF ABS(B-A)<Q THEN GOTO 15Ø
13Ø PRINT "LIMITS";A;"AND";B
14Ø GOTO 9Ø
15Ø PRINT:PRINT"ROOT IS";(A+B)/2
```

If the initial limits are entered as 0,1 this program produces the root after 30 steps, thus:

```
INITIAL LIMITS? Ø,1
LIMITS .5 AND 1
LIMITS .5 AND .75
LIMITS .5 AND .625
LIMITS .5625 AND .625

.........................
.........................

LIMITS .567143261 AND .567143291
LIMITS .567143276 AND .567143291
LIMITS .567143284 AND .567143291
LIMITS .567143288 AND .567143291
LIMITS .567143289 AND .567143291

ROOT IS .567143291
```

The main advantage of the interval bisection method is its predictability. Given the initial limits and the accuracy required we can work out in advance how many steps will be needed. Provided that there is a root between the initial limits, and the function f is continuous between these limits, we can be sure that the root will be found. We pay a price for this reliability: the method is usually very much slower than those described later in this chapter.

An interesting point about the interval bisection method is that we need to know only the sign of $f(c)$, not its value. Although strictly outside the scope of this book, this has an important application in string handling. We can say that a string A$ is earlier in alphabetical order than a string B$ (written as A$ < B$ on most computers), but we cannot quantify the difference B$ – A$. If we have a long list of names stored on the computer in alphabetical order, and we wish to find a particular name in the list, we can use an adaptation of the interval bisection method. We first compare the name required with the name in the middle of the list; we can then confine our search to one-half of the original list, and we compare with the name in the middle of this half; and so on. This technique is known as a **binary search**, and it forms an essential part of much business software.

3.3 Newton's Method

Newton's method—sometimes called the Newton–Raphson method—is the most commonly known method for solving equations of the form (3.1), and many readers will have come across it previously. Some may have spotted that the second-order process (2.10) which we pulled out of a hat in the last chapter was really an illustration of Newton's method.

In Fig. 3.2, we show the graph of $y = f(x)$ which crosses the x-axis at the point R, corresponding to the required root. Our current approximation to the root is x_r, which gives the point P on the curve. We draw the tangent to the curve at P, to cut the axis at T. Provided that the distance PR is reasonably small, the curve should

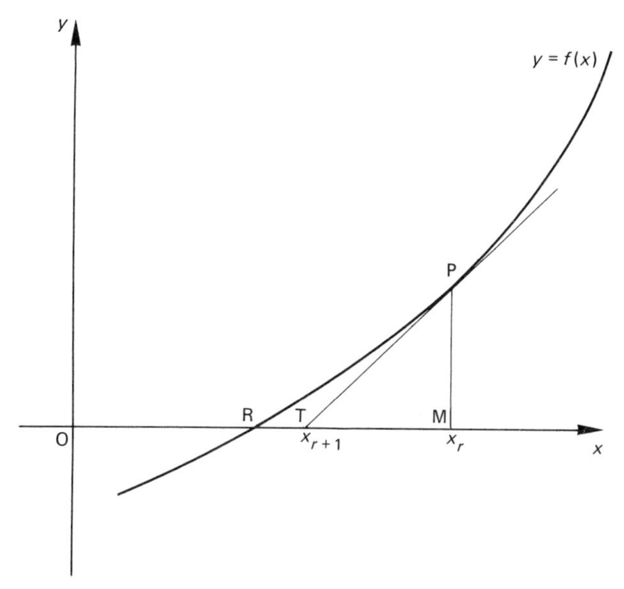

Fig. 3.2 Derivation of Newton's method

not deviate too much from a straight line in this interval, and so T should be quite close to R. We take the position of T as our next approximation to the root, x_{r+1}.

Now the height PM is $f(x_r)$ and tan $\widehat{PTM} = f'(x_r)$, so by simple trigonometry

$$x_{r+1} = x_r - \frac{f(x_r)}{f'(x_r)}. \tag{3.3}$$

This is the formula for Newton's method.

As in Section 2.5, let

$$x_r = \alpha + \varepsilon_r,$$

where α is the correct value of the root. Then by Taylor series

$$f(\alpha) = f(x_r - \varepsilon_r) = f(x_r) - \varepsilon_r f'(x_r) + \tfrac{1}{2}\varepsilon_r^2 f''(x_r) - \dots.$$

But $f(\alpha) = 0$, since α is a root of the equation, and so

$$f(x_r) = \varepsilon_r f'(x_r) - \tfrac{1}{2}\varepsilon_r^2 f''(x_r) + \dots.$$

Thus from equation (3.3)

$$x_{r+1} = \alpha + \varepsilon_r - \varepsilon_r + \tfrac{1}{2}\varepsilon_r^2 \frac{f''(x_r)}{f'(x_r)} - \dots,$$

and it follows that

$$\varepsilon_{r+1} = \tfrac{1}{2}\varepsilon_r^2 \frac{f''(x_r)}{f'(x_r)} \simeq \tfrac{1}{2}\varepsilon_r^2 \frac{f''(\alpha)}{f'(\alpha)},$$

provided that x_r is reasonably close to α. The process is thus seen to be second-order. There is likely to be some difficulty if $f'(x) = 0$ at or near to the required root, for then we have to divide by a number close to zero; we return to this point in Section 3.5.

The following program uses Newton's method to solve equation (3.2); to solve any other equation the expressions for F and G in lines 5Ø and 6Ø must be changed in an obvious way.

```
1Ø REM NEWTON'S METHOD
2Ø INPUT"XØ";X
3Ø LET R=Ø:LET Q=1E-9
4Ø LET R=R+1
5Ø LET F=EXP(-X)-X
6Ø LET G=-EXP(-X)-1
7Ø LET Y=F/G:LET X=X-Y
8Ø PRINT R,X
9Ø IF ABS(Y)>Q THEN GOTO 4Ø
1ØØ PRINT:PRINT "ROOT IS";X
```

The process converges quite rapidly, in six steps at most if x_0 is taken as any positive number. Thus if x_0 is taken as 1 we get the following output:

```
XØ? 1
 1          .537882843
 2          .566986991
 3          .567143286
 4          .567143291
 5          .56714329

ROOT IS .56714329
```

Usually Newton's method converges well and quickly. Its convergence cannot, however, be guaranteed, and it may sometimes converge to a different root from the one expected. In particular, there may be difficulties if there is a point of inflection near the required root. In most serious applications of Newton's method, the process is modified so that $f'(x)$ is not evaluated in every step. Once we are reasonably close to the root, subsequent changes in $f'(x)$ make little difference to the rate of convergence, and some saving in computer time can be achieved by taking advantage of this. You may like to experiment with this idea.

3.4 The Secant Method

The main obstacle to using Newton's method is that it may be difficult or impossible to differentiate the function f. There is no difficulty, of course, if f is defined by a simple mathematical formula. But quite often it is defined only by means of a lengthy computer procedure which is not amenable to differentiation in the ordinary sense of the word. In these circumstances we can use the **secant method**, which is illustrated in Fig. 3.3. If you compare Fig. 3.3 with Fig. 3.2, you will see that the tangent at P in Fig. 3.2 is replaced in Fig. 3.3 by a chord through two points P and Q. If P and Q are close together, the chord should be little different from the tangent.

More formally, P and Q are the points with coordinates $(x_r, f(x_r))$ and $(x_{r-1}, f(x_{r-1}))$. The line through P and Q cuts the x-axis at T, giving the next approximation x_{r+1}. By similar triangles

$$\frac{\text{TM}}{\text{PM}} = \frac{\text{PS}}{\text{QS}} = \frac{x_{r-1} - x_r}{f(x_{r-1}) - f(x_r)}$$

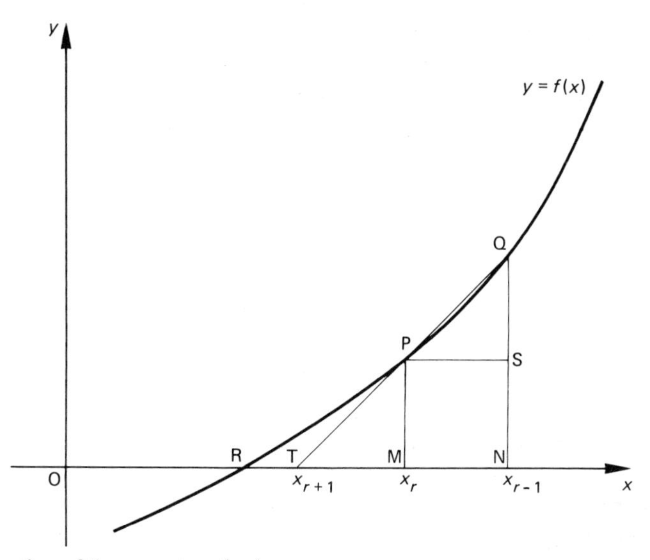

Fig. 3.3 Derivation of the secant method

and so

$$x_{r+1} = x_r - \text{TM} = x_r - \left(\frac{x_{r-1} - x_r}{f(x_{r-1}) - f(x_r)} \right) f(x_r). \tag{3.4}$$

This is the basic formula for the secant method. We need two starting values x_0, x_1 in order to start the process, but the rest is straightforward. A suitable program is given below. Note that X, Y, Z, F, G are used to store the values of x_{r-1}, x_r, x_{r+1}, $f(x_{r-1})$, $f(x_r)$, and these are updated ready for the next step in line 9Ø.

```
1Ø REM SECANT METHOD
2Ø DEF FNF(X)=EXP(-X)-X
3Ø LET Q=1E-9
4Ø INPUT"XØ,X1";X,Y
5Ø LET F=FNF(X):LET G=FNF(Y)
6Ø LET R=1
7Ø LET R=R+1
8Ø LET Z=Y-(X-Y)*G/(F-G)
9Ø LET X=Y:LET Y=Z:LET F=G:LET G=FNF(Y)
1ØØ PRINT R,Y
11Ø IF ABS(Y-X)>Q THEN GOTO 7Ø
12Ø PRINT:PRINT "ROOT IS";Y
```

If the initial values are taken as 0,1 the output is as follows:

```
XØ,X1? Ø,1
 2          .612699837
 3          .563838389
 4          .56717Ø358
 5          .567143307
 6          .56714329
 7          .567143291

ROOT IS .567143291
```

The secant method usually takes a few more steps than Newton's method, but since only $f(x)$ (and not $f'(x)$) has to be evaluated in each step the method can often be more economical in terms of computing time. It suffers from the same disadvantages as Newton's method—that convergence to a particular root cannot be guaranteed—but nevertheless it is a powerful general purpose method.

Another method sometimes used is the **rule of false position**. This is very similar to the secant method, but for this method the two points P and Q in Fig. 3.3 are always on opposite sides of the root R. We start with two values a, b such that $f(a)$, $f(b)$ are of opposite signs. We then take

$$c = b - \left(\frac{a - b}{f(a) - f(b)} \right) f(b),$$

which is derived in the same way as equation (3.4). If $f(c)$ has the same sign as $f(a)$ we replace a by c; otherwise we replace b by c. We repeat this until a and b are sufficiently close together to give the accuracy required. By this stage, you should be able to write your own program to implement this method. The rule of false position can normally be relied upon to converge to the root required, but sometimes its convergence is exceedingly slow. It can be modified to improve its rate of convergence, but we shall not pursue this here.

3.5 Multiple Roots

We shall now apply Newton's method to the equation

$$x^2 - 3x + 2.25 = 0. \tag{3.5}$$

To do this, we simply change lines 5Ø and 6Ø of the program in Section 3.3 to read:

```
5Ø LET F=X*X-3*X+2.25
6Ø LET G=2*X-3
```

If we set $x_0 = 1$, the process converges after 16 steps to the value of 1.499 984 74; if we set $x_0 = 2$, the process converges after 16 steps to the value 1.500 015 26. We might therefore imagine that the equation has two roots given by these two values.

But in fact equation (3.5) factorises into

$$(x - 1.5)^2 = 0,$$

and consequently has a double root equal to 1.5. Both the above values are approximations to 1.5: but they are accurate only to four decimal places, not eight as we might expect. So what has gone wrong? In this problem, $f(x) = (x - 1.5)^2$. If $x = 1.5 + \varepsilon$, then $f(x) = \varepsilon^2$; a small change in x gives a much smaller change in $f(x)$. In fact, for any value of x between 1.499 984 74 and 1.500 015 26, the computer evaluates $f(x)$ as zero; so it accepts any value in this region as a root of the equation.

Generally speaking, when we evaluate a double root, we obtain only about half the accuracy of which the computer is capable. If we *know* that an equation has a double root, then we can easily overcome this difficulty, for a double root of $f(x) = 0$ is a single root of $f'(x) = 0$, and the latter equation presents no problems. But usually we do not know in advance that an equation has a double root, although we might suspect it from our sketch graph. We should always be suspicious if we obtain two roots close together as in the above example; even if there really are two distinct roots we have probably suffered some loss of accuracy. As you might expect, the situation is even worse for multiple roots of order higher than two.

Most problems which lead to multiple roots, are in fact ill-posed problems, i.e. problems which have no sensible numerical solutions. Suppose that in equation (3.5) the number 2.25 is rounded to two decimal places: that is, it represents a number somewhere between 2.245 and 2.255. The equation

$$x^2 - 3x + 2.245 = 0$$

has the pair of real roots 1.4293, 1.5707, but the equation

$$x^2 - 3x + 2.255 = 0$$

has no real roots. So in these circumstances we can say nothing useful about the roots of equation (3.5). It is the problem itself which is at fault, not our methods for dealing with it.

3.6 Polynomial Equations

Polynomial equations of the form

$$a_0x^n + a_1x^{n-1} + a_2x^{n-2} + \ldots + a_{n-1}x + a_n = 0 \tag{3.6}$$

occur frequently in many branches of mathematics, the humble quadratic equation being the most obvious example. Equation (3.6) has exactly n roots, but some may be complex and some may occur as multiple roots. We shall consider here only the evaluation of real roots. We shall in fact use Newton's method to obtain these roots, and what we shall be discussing in this section are the technicalities of implementing Newton's method for these problems.

First, let us consider the evaluation of a polynomial on a computer. The most obvious way to evaluate the polynomial

$$2x^4 + 3x^3 + 4x^2 + 5x + 6$$

at a value $x = X$ is to use the instruction

```
LET F=2*X↑4+3*X↑3+4*X↑2+5*X+6                                   (3.7)
```

but in fact it is more efficient to write

```
LET F=(((2*X+3)*X+4)*X+5)*X+6                                   (3.8)
```

The instructions (3.7) and (3.8) both require four additions and four multiplications, but (3.7) requires a further three exponentiations (↑). Exponentiations are slow compared with other arithmetical operations, and the use of the form (3.8)—known as **nested multiplication**—can result in a considerable saving in computer time if the polynomial is evaluated frequently.

Suppose now we wish to evaluate the polynomial

$$f(x) = a_0x^n + a_1x^{n-1} + \ldots + a_{n-1}x + a_n$$

at a value $x = X$, and we want our program to work for any value of n. We cannot use instruction (3.8) directly: for a start, we would not know how many brackets to put at the beginning! But we can use the same idea by writing

$$\begin{aligned} b_0 &= a_0 \\ b_1 &= b_0X + a_1 = a_0X + a_1 \\ b_2 &= b_1X + a_2 = (a_0X + a_1)X + a_2 \\ b_3 &= b_2X + a_3 = ((a_0X + a_1)X + a_2)X + a_3 \end{aligned}$$

and in general

$$b_r = b_{r-1}X + a_r, \quad r = 1, 2, \ldots, n. \tag{3.9}$$

The value of b_n is then equal to $f(X)$. Try this for a simple polynomial with a simple value of X, just to convince yourself that it is correct.

We can do some surprising things with polynomials, and you may find the next result particularly surprising. Define a new polynomial $g(x)$ of degree $n-1$ by

$$g(x) = b_0 x^{n-1} + b_1 x^{n-2} + \dots + b_{n-2} x + b_{n-1}. \tag{3.10}$$

Then

$$\begin{aligned}(x-X)g(x) \\ &= b_0 x^n + (b_1 - b_0 X)x^{n-1} + (b_2 - b_1 X)x^{n-2} + \dots + (b_{n-1} - b_{n-2}X)x - b_{n-1}X \\ &= a_0 x^n + a_1 x^{n-1} + a_2 x^{n-2} + \dots + a_{n-1}x + (a_n - b_n) \\ &= f(x) - b_n,\end{aligned}$$

by virtue of equation (3.9). Thus we may write

$$f(x) = (x-X)g(x) + b_n, \tag{3.11}$$

which incidentally confirms that $f(X) = b_n$. It follows from equation (3.11) that

$$f'(x) = g(x) + (x-X)g'(x)$$

and hence that

$$f'(X) = g(X).$$

To evaluate $f'(X)$, therefore, we simply apply the nested multiplication technique to $g(X)$. That is we set

$$c_0 = b_0$$

$$c_r = c_{r-1}X + b_r, \quad r = 1, 2, \dots, n-1,$$

and then $f'(X) = g(X) = c_{n-1}$. The above procedures give us an efficient method for evaluating f and f' in Newton's method.

If it so happens that the equation $f(x) = 0$ has a root at $x = X$, then b_n will be zero, at least to the accuracy with which we are working. In this case we can see from equation (3.11) that $f(x)$ can be factorised into $(x - X)$ and $g(x)$. The remaining roots of $f(x) = 0$ are the roots of $g(x) = 0$. When we have found the first root of the equation, we replace $f(x)$ by $g(x)$ before proceeding to the next root, thus reducing the degree of the polynomial by one. We continue in this way until we have found all the real roots.

The program below implements these ideas. This program is longer and more complicated than those we have used hitherto, so we have divided it into blocks separated by REM statements. You may please yourself whether or not you type these statements into the computer. What you really should do—and this applies to all the programs in this book—is to insert your own REM statements wherever you feel an explanation is necessary for your own purposes. In the program the arrays A(R) and B(R) store the values of a_r and b_r; c_r is stored simply as C, since we never need more than one c_r to be stored at a time. Note the instruction in line 21Ø which deals with the trivial case $n = 1$ without resorting to Newton's method. Note also that in finding the first root, x_0 is taken as zero; for subsequent roots x_0 is taken equal to the root just found.

```
10 REM POLYNOMIAL EQUATIONS
20 REM ***** INPUT DATA *****
30 INPUT"DEGREE OF POLYNOMIAL";N
40 DIM A(N),B(N):PRINT
50 FOR R=0 TO N
60 PRINT "A(";R;") =";:INPUT A(R)
70 NEXT R
80 PRINT:INPUT"NUMBER OF REAL ROOTS";M
90 LET X=0:LET Q=1E-9:PRINT
100 REM ***** MAIN LOOP *****
110 FOR S=1 TO M
200 REM ***** DEAL WITH N=1 *****
210 IF N=1 THEN LET X=-A(1)/A(0):GOTO 530
300 REM ***** EVALUATE F(X) *****
310 LET B(0)=A(0)
320 FOR R=1 TO N
330 LET B(R)=B(R-1)*X+A(R)
340 NEXT R
400 REM ***** EVALUATE F'(X)*****
410 LET C=B(0)
420 FOR R=1 TO N-1
430 LET C=C*X+B(R)
440 NEXT R
500 REM ***** NEWTON FORMULA *****
510 LET Y=B(N)/C:LET X=X-Y
520 IF ABS(Y)>Q THEN GOTO 320
530 PRINT "ROOT NO.";S;"=";X
600 REM ***** PREPARE FOR NEXT ROOT *****
610 IF S=M THEN GOTO 710
620 LET N=N-1
630 FOR R=1 TO N:LET A(R)=B(R):NEXT R
700 REM ***** END OF MAIN LOOP *****
710 NEXT S
```

When solving the equation

$$5x^4 - 24x^3 + 22x^2 - 8x + 1 = 0,$$

this gives the output:

```
DEGREE OF POLYNOMIAL? 4

A( 0 ) =? 5
A( 1 ) =? -24
A( 2 ) =? 22
A( 3 ) =? -8
A( 4 ) =? 1

NUMBER OF REAL ROOTS? 2

ROOT NO. 1 = .267949193
ROOT NO. 2 = 3.73205081
```

Remember that Newton's method does not converge if there are no real roots. If we ask for too many real roots the computer will go into an endless iteration, which can only be stopped by pressing the STOP or BREAK key. Try this by asking for three real roots with the above equation. You might like to modify the program so that it automatically stops with a suitable message if it has found no root after (say) fifty iterations.

Exercises

1 A list of N names are stored in the computer in alphabetical order as an array A$(1), A$(2), ... A$(N). Write a subroutine to find a specified name B$ in this list, using the binary search method described in Section 3.2.

2 Show graphically that the equation

$$4 \sin x = 1 + x$$

has three real roots. Find all of them using (i) the interval bisection method, (ii) Newton's method, (iii) the secant method.

3 Show that the equation

$$\tan x = 5x$$

has an infinite number of roots.

(i) Find the smallest positive root using Newton's method.
(ii) Use Newton's method with various starting values (e.g. 4, 5, 4.4, 4.5, 4.6, 4.7) to find the second smallest positive root. Comment on your results.
(iii) Find the third smallest positive root.

4 Write a program to implement the rule of false position described in Section 3.4. Use it to solve the equation

$$e^{-x} = x,$$

starting with the pairs of values (0,1), (0,100), (−10,1). Compare with the secant method using the same starting values.

5 Use the program for polynomial equations to find the three real roots of the equation

$$x^3 - 9x^2 + 26.25x - 24.5 = 0.$$

How do your answers compare with the theoretical roots of this equation?

6 Show that the iterative process

$$x_{r+1} = x_r - \frac{f(x_r)}{f'(x_r)} - \frac{\{f(x_r)\}^2 f''(x_r)}{2\{f'(x_r)\}^3}$$

is a third-order process for solving the equation $f(x) = 0$. Use it to solve the equation in question 2.

4

Simultaneous Equations I

4.1 Simple Elimination Techniques

Most of us learnt how to solve simultaneous equations at an early stage of our mathematical education. Thus to solve the pair of equations

$$2x + y = 7$$

$$6x + 5y = 27,$$

we first eliminate x: i.e. we multiply the first equation by 3 and subtract from the second, to give

$$2y = 6.$$

We thus have $y = 3$, and by substitution into the first equation we immediately find that $x = 2$. You may be surprised to find that the methods used on computers to solve large sets of simultaneous equations are essentially based on the elimination technique we have just used.

In order to illustrate these elimination techniques—usually known as **Gaussian elimination**—in a slightly more difficult situation, we shall now work through the solution of four equations in four unknowns. We shall then be in a position to write a computer program to solve n equations in n unknowns, where n is restricted only by the size of the computer's memory.

Consider the set of equations:

$$\left.\begin{aligned}
&(1) \quad 2w + 4x + y + 2z = 5 \\
&(2) \quad 4w + 14x - y + 6z = 11 \\
&(3) \quad w - x + 5y - z = 9 \\
&(4) \quad -4w + 2x - 6y + z = -2
\end{aligned}\right\} \quad (4.1)$$

Our first task is to eliminate the unknown w. We do this by subtracting an appropriate multiple of equation (1) from each of equations (2), (3), (4) to give equations (5), (6), (7) below:

$$\begin{aligned}
&(5) \quad 6x - 3y + 2z = 1 && (2) - 2 \times (1) \\
&(6) \quad -3x + \tfrac{9}{2}y - 2z = \tfrac{13}{2} && (3) - \tfrac{1}{2} \times (1) \\
&(7) \quad 10x - 4y + 5z = 8 && (4) + 2 \times (1)
\end{aligned}$$

Note that we have used equation (1) in a rather different way from the other three equations, and we call this the **pivotal equation**.

We have now reduced the problem to one of three equations in three unknowns. By repeating the procedure we get two equations in two unknowns, then one equation in one unknown, which we can solve immediately. If we are doing the work by hand (and thanks to modern technology this is something you will very rarely need to do!) It is convenient to set the work out in tabular form as shown in Table 4.1. Notice that in this table we have introduced an extra column headed SUM, which is used as a check on our arithmetic—something we can dispense with when we put the problem on the computer. In lines (1)–(4) this column is simply the sum of the other five numbers on the line. For the remaining lines, the sum is worked out in exactly the same way as the other numbers on the line: $34 - 2 \times 14 = 6$, $13 - \frac{1}{2} \times 14 = 6$, and so on. If our arithmetic is correct, the sum column should be equal to the sum of the other terms on the line: thus $6 - 3 + 2 + 1 = 6$, etc.

Lines (8) and (9) in the table show the result of eliminating the unknown x from equations (5), (6) and (7); whilst line (10) gives the result of eliminating y from equations (8) and (9). This last equation immediately gives us $z = 2$.

Table 4.1 Solution of equations (4.1)

	w	x	y	z	RHS	SUM	
* (1)	2	4	1	2	5	14	
(2)	4	14	−1	6	11	34	
(3)	1	−1	5	−1	9	13	
(4)	−4	2	−6	1	−2	−9	
* (5)		6	−3	2	1	6	$(2) - 2 \times (1)$
(6)		−3	$\frac{9}{2}$	−2	$\frac{13}{2}$	6	$(3) - \frac{1}{2} \times (1)$
(7)		10	−4	5	8	19	$(4) + 2 \times (1)$
* (8)			3	−1	7	9	$(6) + \frac{1}{2} \times (5)$
(9)			1	$\frac{5}{3}$	$\frac{19}{3}$	9	$(7) - \frac{5}{3} \times (5)$
(10)				2	4	6	$(9) - \frac{1}{3} \times (8)$

We now use the technique of **back substitution** in the pivotal equations (marked with an asterisk in the table) to find y, x, w in turn. Thus equations (8), (5), (1) give

$$y = \tfrac{1}{3}(7 + z) = 3,$$

$$x = \tfrac{1}{6}(1 + 3y - 2z) = 1,$$

$$w = \tfrac{1}{2}(5 - 4x - y - 2z) = -3.$$

This completes the solution of the equations.

Note that when we do the calculation on a computer, we do not need to find room in the store for all the numbers in Table 4.1. Once we have obtained line (5) we no longer need line (2); so we can store line (5) in the space previously occupied by line (2). Similarly, lines (6), (7), (8), (9), (10) can overwrite lines (3), (4), (6), (7), (9) respectively. At the end of the elimination process, only lines (1), (5), (8) and (10) are held in the computer memory. Provided that there is enough room in the memory to hold the original equations, there is enough room to solve them; there should be no danger of a frustrating OUT OF MEMORY ERROR in the middle of the computation.

4.2 Computer Implementation

The general problem is to solve the set of equations

$$
\begin{array}{ll}
(0) & a_{00}x_0 + a_{01}x_1 + a_{02}x_2 + \ldots + a_{0n}x_n = b_0 \\
(1) & a_{10}x_0 + a_{11}x_1 + a_{12}x_2 + \ldots + a_{1n}x_n = b_1 \\
 & \ldots \quad \ldots \quad \ldots \quad \ldots \quad \ldots \\
(n) & a_{n0}x_0 + a_{n1}x_1 + a_{n2}x_2 + \ldots + a_{nn}x_n = b_n.
\end{array}
$$

We have started our numbering at 0 rather than 1 because on most microcomputers arrays start at 0. We shall store a_{rs} as an array A(R, S), with R and S going from 0 to N. We shall also store b_r in the same array as A(R, N + 1), but we shall use a separate array X(R) to hold the unknowns x_r. In the present context it is appropriate to think of an array as holding a matrix or a vector; indeed the Pure Mathematician would prefer to write the above set of equations in matrix form as

$$\boldsymbol{Ax} = \boldsymbol{b}.$$

We shall make use of this matrix notation in the next chapter, but for the moment it is not particularly helpful.

For the first stage in the solution of the equations, we take line (0) as the pivotal equation. This line is left unchanged, but for $r = 1, 2, \ldots n$ we replace line (r) by

$$(r) - \frac{a_{r0}}{a_{00}} \times (0).$$

The equations now look like this:

$$
\begin{array}{ll}
(0) & a_{00}x_0 + a_{01}x_1 + a_{02}x_2 + \ldots + a_{0n}x_n = b_0 \\
(1) & \qquad a'_{11}x_1 + a'_{12}x_2 + \ldots + a'_{1n}x_n = b'_1 \\
 & \qquad \ldots \quad \ldots \quad \ldots \quad \ldots \\
(n) & \qquad a'_{n1}x_1 + a'_{n2}x_2 + \ldots + a'_{nn}x_n = b'_n
\end{array}
$$

The value of a'_{11} (for instance) is different from the value of a_{11}, but it is stored in the same place in the computer.

The next stage in the solution is the elimination of x_1 from lines (2) to (n). In this stage lines (0) and (1) are unaltered, but for $r = 2, 3, \ldots n$, line (r) is replaced by

$$(r) - \frac{a'_{r1}}{a'_{11}} \times (1).$$

We continue in the same way until we have eliminated all the unknowns except x_n.

We give below a program for solving a general set of equations by the Gaussian elimination procedure:

```
10 REM GAUSSIAN ELIMINATION
20 REM ***** INPUT DATA *****
30 INPUT"NUMBER OF EQUATIONS";N
40 LET N=N-1:PRINT
50 DIM A(N,N+1),X(N)
60 FOR R=0 TO N
70 FOR S=0 TO N
80 PRINT "A(";R;",";S;") =";
90 INPUT A(R,S)
100 NEXT S
110 PRINT:PRINT "B(";R;") =";
120 INPUT A(R,N+1)
130 PRINT:PRINT:NEXT R
200 REM ***** ELIMINATION *****
210 FOR Z=0 TO N-1
400 FOR R=Z+1 TO N
410 LET P=A(R,Z)/A(Z,Z)
420 FOR S=Z+1 TO N+1
430 LET A(R,S)=A(R,S)-P*A(Z,S)
440 NEXT S
450 NEXT R
500 NEXT Z
600 REM ***** BACK SUBSTITUTION *****
610 FOR R=N TO 0 STEP -1
620 LET P=A(R,N+1)
630 IF R=N THEN GOTO 670
640 FOR S=R+1 TO N
650 LET P=P-A(R,S)*X(S)
660 NEXT S
670 LET X(R)=P/A(R,R)
680 NEXT R
700 REM ***** PRINT RESULTS *****
710 FOR R=0 TO N
720 PRINT"X(";R;") =";X(R)
730 NEXT R
```

In this program, lines 30–130 are simply an input routine to get the equations into the computer. The elimination procedure is contained in lines 210–500. The variable Z stores the number of the unknown currently being eliminated, and goes through the values 0, 1, . . . N − 1. The back substitution is carried out in lines 610–680, and the remainder of the program simply prints out the results. Make sure that you really understand what is happening in each part of this program.

You may have thought hitherto that the solution of a large set of simultaneous equations was a very complicated piece of computation. Yet this is not a long program, and roughly half of it is taken up with input and output routines. It will work for any value of N, subject only to the size of your computer's memory: but if N is large the computer will be kept busy for quite a long time!

We can test the program using the set of equations (4.1); this gives the expected output:

```
X( 0 ) =-3
X( 1 ) = 1
X( 2 ) = 3
X( 3 ) = 2
```

4.3 Ill-conditioning

If you try the above program with the pair of equations

$$\left.\begin{aligned} x + y &= 1 \\ 2x + 2y &= 3 \end{aligned}\right\} \quad (4.2)$$

or

$$\left.\begin{aligned} x + y &= 1 \\ 2x + 2y &= 2 \end{aligned}\right\} \quad (4.3)$$

you will get

```
?DIVISION BY ZERO ERROR IN 670
```

This should not surprise you. Both these sets of equations are *singular*, that is the determinant of the coefficients on the left-hand side is zero. Clearly equations (4.2) have no possible solution, and equations (4.3) have an infinite number of solutions. It would be most surprising if our program did not produce an error message.

In numerical work, we are not so much bothered by singular equations as by equations which are nearly, but not quite, singular. We have already met a problem like this in Exercise 6 of Chapter 1, namely

$$\left.\begin{aligned} 7x + 2y &= 6 \\ x + 0.28y &= 2. \end{aligned}\right\} \quad (4.4)$$

If we solve these using the above program, we get the answers:

```
X( 0 ) = 58.0000004
X( 1 ) =-200.000002
```

But in fact if 0.28 is interpreted as a number between 0.275 and 0.285, x could have any value between 31.3 and 458.0, and y any value between -1600.0 and -106.7. If we changed 0.28 to $\frac{2}{7} = 0.285\,714\ldots$, the equations would be singular and have no solution. Equations such as these are said to be *ill-conditioned*, and as we have just seen there can be a substantial loss of accuracy in their solution. The fault lies with the problem, not with the method of solution, but we must be careful to ensure that our method of solution does not aggravate the ill-conditioning. It is by no means easy to recognise an ill-conditioned set of equations, except in the case $n = 2$. It is quite likely that at some stage of your career you will accept solutions as accurate when they are no more meaningful than those of equations (4.4); at least you have been warned!

4.4 The Need for Pivoting

Now try solving the following set of equations using the program from Section 4.2:

$$\left.\begin{aligned} x + y + 2z &= 9 \\ x + y + 3z &= 12 \\ 2x + 3y + z &= 11. \end{aligned}\right\} \quad (4.5)$$

These equations are not singular, nor even ill-conditioned, yet the program gives

```
?DIVISION BY ZERO ERROR IN 410
```

So what has gone wrong here? We can see this quite easily by working through the calculation by hand, as in Table 4.2. The next step in this table should be to take $(5) - p \times (4)$ so as to eliminate y. The required value of p is 1/0, and hence the DIVISION BY ZERO ERROR. The sensible thing to do at this stage would be to interchange lines (4) and (5): but computers are not sensible, they merely do what we tell them to do.

Table 4.2 Solution of equations (4.5)

	x	y	z	RHS	SUM	
* (1)	1	1	2	9	13	
(2)	1	1	3	12	17	
(3)	2	3	1	11	17	
* (4)		0	1	3	4	(2) – 1 × (1)
(5)		1	–3	–7	–9	(3) – 2 × (1)

A potentially more dangerous situation arises if we solve the following equations:

$$\left.\begin{aligned} x + y + 2z &= 9 \\ x + 1.000\,000\,1y + 3z &= 12.000\,000\,2 \\ 2x + 3y + z &= 11. \end{aligned}\right\} \qquad (4.6)$$

These equations have the same solutions ($x = 1$, $y = 2$, $z = 3$) as equations (4.5), but the solution given by the PET computer is:

```
X( Ø ) =-9.34559666E-Ø3
X( 1 ) = 3.ØØ93458
X( 2 ) = 2.9999999
```

And we have no error message to warn us that something is amiss! If we again carry out the computation by hand, the zero in line (4) of Table 4.2 becomes 0.000 000 1, obtained by subtracting 1 from 1.000 000 1. Since we have subtracted two nearly equal numbers we have lost most of the accuracy, and from this point onwards the computation goes haywire. If we continued the calculation by hand we should in fact get the correct answers; since we work in decimal, we can store the number 0.000 000 1 exactly, and there is no loss of accuracy.

If you look back to Sections 4.1 and 4.2, you will find that the only numbers we need to divide by are the first coefficients in the pivotal equations: for instance the numbers 2, 6, 3 in Table 4.1. These numbers are called **pivots**, and it is important that the pivots should not be zero or nearly zero. We can usually avoid small pivots by changing the order of the equations, for instance by interchanging lines (4) and (5) in Table 4.2. Any serious program for simultaneous equations must include provision for rearrangements of this kind, usually known as **pivoting**.

4.5 Partial Pivoting

Lines 4ØØ–45Ø of the program in Section 4.2 carry out the elimination of the unknown X(Z) from the set of equations defined by

(Z)	A(Z, Z)	A(Z, Z + 1)	...	A(Z, N)	A(Z, N + 1)
(Z + 1)	A(Z + 1, Z)	A(Z + 1, Z + 1)	...	A(Z + 1, N)	A(Z + 1, N + 1)
	...	...		...	...
(N)	A(N, Z)	A(N, Z + 1)	...	A(N, N)	A(N, N + 1)

In line 41Ø we divide by the pivot A(Z, Z), and we want to ensure that this pivot is not too small. What we do is to look through the numbers in the first column of the above table—A(Z, Z), A(Z + 1, Z), ..., A(N, Z)—and pick out the one which is largest in magnitude, say A(U, Z). We then interchange lines (Z) and (U) and proceed as before. All we are doing is changing the order in which the equations are written down, and this in no way affects the solution of the equations. We call this technique **partial pivoting.**

In order to implement this technique, we first change line 1Ø of the previous program to

```
1Ø REM GAUSSIAN ELIMINATION - PARTIAL PIVOTING
```

and then insert the lines:

```
22Ø LET W=Ø
23Ø FOR R=Z TO N
24Ø IF ABS(A(R,Z))>W THEN LET U=R:LET W=ABS(A(R,Z))
25Ø NEXT R
26Ø IF U=Z THEN GOTO 4ØØ
27Ø FOR S=Z TO N+1
28Ø LET P=A(U,S)
29Ø LET A(U,S)=A(Z,S)
3ØØ LET A(Z,S)=P
31Ø NEXT S
```

(Now you can see why we left some large gaps in the original line numbering!) Lines 22Ø–25Ø use a standard method for finding the maximum of a set of numbers, and lines 27Ø–31Ø interchange the two rows. Line 26Ø avoids the interchange if U = Z, although the program will still work if this line is omitted.

If we use this amended program for the sets of equations (4.5) and (4.6) we get the acceptable results

```
X( Ø ) = 1
X( 1 ) = 2
X( 2 ) = 3
```

and

```
X( Ø ) = .999999998
X( 1 ) = 2
X( 2 ) = 3
```

Problem solved!

4.6 Total Pivoting

... or at least, almost solved. In fact the program just developed is perfectly adequate for most sets of simultaneous equations which arise in practice, but if we try hard we can find sets of equations which fool it. The set of equations

$$\left.\begin{aligned} 2x + 2y + 4z &= 18 \\ x + 1.000\,000\,1y + 3z &= 12.000\,000\,2 \\ 0.000\,000\,02x + 0.000\,000\,03y + 0.000\,000\,01z &= 0.000\,000\,11 \end{aligned}\right\} \quad (4.7)$$

is obtained from the set (4.6) by multiplying the first equation by 2 and the third equation by 0.000 000 01. This does not, of course, affect the solutions, but if we use the partial pivoting program we get the incorrect answers:

```
X( Ø ) =-9.34559666E-Ø3
X( 1 ) = 3.ØØ93458
X( 2 ) = 2.9999999
```

For these equations, after eliminating the variable x we are left with the table:

$$\begin{array}{cc|c} 0.000\,000\,1 & 1 & 3.000\,000\,2 \\ 0.000\,000\,01 & -0.000\,000\,03 & -0.000\,000\,07 \end{array}$$

The term 0.000 000 1 is obtained as the difference between two nearly equal numbers, and is highly inaccurate; but the program chooses this number as the pivot in preference to the smaller but accurate 0.000 000 01.

We can usually overcome problems like this by using **total pivoting**. We search for the largest number (in magnitude) in the entire array

$$\begin{array}{cccc} A(Z, Z) & A(Z, Z + 1) & \ldots & A(Z, N) \\ A(Z + 1, Z) & A(Z + 1, Z + 1) & \ldots & A(Z + 1, N) \\ \ldots & \ldots & & \ldots \\ A(N, Z) & A(N, Z + 1) & \ldots & A(N, N) \end{array}$$

instead of just in the first column, and make this number the pivot. This means that we shall probably need to interchange two columns as well as two rows; and if we interchange two columns we change the order in which the unknowns are stored. We must somehow keep a record of all the column interchanges, or when we come to the back substitution we shall not know which unknown is which.

We introduce an additional array C(S), where C(S) is the number of the unknown currently held in the Sth column: i.e. the Sth column holds the unknown X(C(S)). Whenever we do a column interchange, we amend C(S) appropriately. In order to develop a total pivoting program, we make the following changes to the program of Section 4.2:

(i) Change line 1Ø to

```
1Ø REM GAUSSIAN ELIMINATION - TOTAL PIVOTING
```

(ii) Insert the following lines, to initialise the array C(S):

```
15Ø DIM C(N)
16Ø FOR S=Ø TO N
17Ø LET C(S)=S
18Ø NEXT S
```

(iii) Insert these additional lines to carry out the pivoting:

```
22Ø LET W=Ø
23Ø FOR R=Z TO N:FOR S=Z TO N
24Ø IF ABS(A(R,S))>W THEN LET U=R:LET V=S:LET W=ABS(A(R,S))
25Ø NEXT S:NEXT R
26Ø IF U=Z THEN GOTO 32Ø
27Ø FOR S=Z TO N+1
28Ø LET P=A(U,S)
29Ø LET A(U,S)=A(Z,S)
3ØØ LET A(Z,S)=P
31Ø NEXT S
32Ø IF V=Z THEN GOTO 4ØØ
33Ø FOR R=Ø TO N
34Ø LET P=A(R,V)
35Ø LET A(R,V)=A(R,Z)
36Ø LET A(R,Z)=P
37Ø NEXT R
38Ø LET P=C(V):LET C(V)=C(Z):LET C(Z)=P
```

(iv) Amend the following two lines in the back substitution:

```
65Ø LET P=P-A(R,S)*X(C(S))

67Ø LET X(C(R))=P/A(R,R)
```

Lines 22Ø–25Ø find the largest number A(U, V) in the array, lines 26Ø–31Ø perform the row interchange, and lines 32Ø–37Ø perform the column interchange. Note that when making a column interchange, we change all the terms from $R = 0$ to $R = N$, not just those from Z to N, otherwise we shall go adrift in the back substitution. Line 38Ø brings the array C(S) up to date. In the back substitution we have to use X(C(.)) instead of X(.). You have probably found all this quite confusing! Certainly there are a number of quite tricky points in this program, but please try and understand what is happening. You may find it helpful to insert some additional REM statements for future reference. (If it is any comfort, it took the author a long time to get this program right!)

For the set of equations (4.7), this revised program gives the almost correct answers:

```
X( Ø ) = .999999993
X( 1 ) = 2.ØØØØØØØ1
X( 2 ) = 3
```

Total pivoting takes much more time than partial pivoting, and for most practical purposes partial pivoting should be adequate; but when in doubt, use total pivoting. No amount of pivoting will remove inherent ill-conditioning from a set of equations, but it helps to ensure that no further ill-conditioning is introduced in the course of the computation.

Exercises

1 Modify the input routine of the programs in this chapter so that the coefficients a_{rs} are taken as random numbers, and so that

$$b_r = a_{r0} + 2a_{r1} + 3a_{r2} + \ldots + (n+1)a_{rn}.$$

Run the programs with various values of n and check that you get the correct answers. Find how large a set of equations you can get on your computer, and how long the computer takes to solve them.

2 Using the programs from Question 1, find, for various values of n, the time taken to solve a set of n equations using no pivoting, partial pivoting and total pivoting. Display your results by means of graphs. Count only the time required for the actual solution of the equations, excluding the input and output routines. By suitably amending the programs, you should be able to make the computer do the actual timing.

3 We sometimes need to solve equations of the form

$$AX = B$$

where A is an $n \times n$ matrix, X is an unknown $n \times m$ matrix, and B is a given $n \times m$ matrix. Simultaneous equations correspond to the case $m = 1$. Write a program, using partial pivoting, to solve an equation of this type for any values of m and n. Test your program with

$$A = \begin{bmatrix} 1 & 2 & 3 & 4 \\ 2 & 1 & 4 & 3 \\ 2 & 5 & 7 & 1 \\ 1 & -1 & 3 & 2 \end{bmatrix} \quad \text{and} \quad B = \begin{bmatrix} 20 & 10 & 6 \\ 22 & 10 & 7 \\ 38 & 15 & 5 \\ 9 & 5 & 4 \end{bmatrix}.$$

4 In the last problem we actually evaluated $X = A^{-1}B$, although we did not obtain A^{-1} explicitly. In practice, it is rarely necessary to find an inverse matrix, but we could obtain A^{-1} quite easily by setting B equal to the identity matrix. Use this approach to find the inverse of the matrix

$$\begin{bmatrix} 1 & 2 & 3 & 4 \\ 2 & 1 & 4 & 2 \\ 2 & 5 & 7 & 12 \\ 1 & -1 & 3 & 11 \end{bmatrix}.$$

5

Simultaneous Equations II

5.1 Triangular Factorisation

We start this chapter by taking another look at the solution of equations (4.1). We can write these equations in matrix form as

$$A\boldsymbol{x} = \boldsymbol{b}, \tag{5.1}$$

where

$$A = \begin{bmatrix} 2 & 4 & 1 & 2 \\ 4 & 14 & -1 & 6 \\ 1 & -1 & 5 & -1 \\ -4 & 2 & -6 & 1 \end{bmatrix}, \quad \boldsymbol{x} = \begin{bmatrix} w \\ x \\ y \\ z \end{bmatrix}, \quad \boldsymbol{b} = \begin{bmatrix} 5 \\ 11 \\ 9 \\ -2 \end{bmatrix}.$$

When we have completed the elimination procedure, we are left with lines (1), (5), (8), (10) of Table 4.1. These can be written as

$$U\boldsymbol{x} = \boldsymbol{c}, \tag{5.2}$$

where

$$U = \begin{bmatrix} 2 & 4 & 1 & 2 \\ 0 & 6 & -3 & 2 \\ 0 & 0 & 3 & -1 \\ 0 & 0 & 0 & 2 \end{bmatrix} \quad \text{and} \quad \boldsymbol{c} = \begin{bmatrix} 5 \\ 1 \\ 7 \\ 4 \end{bmatrix}.$$

The matrix U is said to be an **upper triangular** matrix, for obvious reasons. The whole aim of the elimination process is to reduce an equation of the form (5.1) to the form (5.2). Provided that U is upper triangular, this last set of equations is easily solved by back substitution.

We are now going to look more closely at the relationships between lines (1), (2), (3), (4) of Table 4.1, which give equations (5.1), and lines (1), (5), (8), (10), which give equations (5.2). From the last column of the table we can see that

$$\begin{aligned}
(5) &= (2) - 2 \times (1), \\
(8) &= (6) + \tfrac{1}{2} \times (5) = (3) - \tfrac{1}{2} \times (1) + \tfrac{1}{2} \times (5), \\
(10) &= (9) - \tfrac{1}{3} \times (8) = (7) - \tfrac{5}{3} \times (5) - \tfrac{1}{3} \times (8) \\
&= (4) + 2 \times (1) - \tfrac{5}{3} \times (5) - \tfrac{1}{3} \times (8).
\end{aligned}$$

We can thus write

$$\left.\begin{aligned}
(1) &= (1), \\
(2) &= 2 \times (1) + (5), \\
(3) &= \tfrac{1}{2} \times (1) - \tfrac{1}{2} \times (5) + (8), \\
(4) &= -2 \times (1) + \tfrac{5}{3} \times (5) + \tfrac{1}{3} \times (8) + (10).
\end{aligned}\right\} \qquad (5.3)$$

Equations (5.3) are relationships between the rows of A and $\boldsymbol{b}$ and the rows of U and $\boldsymbol{c}$. We can write these relationships more conveniently in terms of matrix multiplication as

$$A = LU, \quad \boldsymbol{b} = L\boldsymbol{c}, \qquad (5.4)$$

where L is the **lower triangular** matrix

$$L = \begin{bmatrix} 1 & 0 & 0 & 0 \\ 2 & 1 & 0 & 0 \\ \frac{1}{2} & -\frac{1}{2} & 1 & 0 \\ -2 & \frac{5}{3} & \frac{1}{3} & 1 \end{bmatrix}.$$

You should check that equations (5.4) are correct by carrying out the matrix multiplication.

The elimination part of the solution of a set of simultaneous equations can now be seen in a different light. What we have done is to **factorise** the matrix A into the product LU, where L is a lower triangular matrix with 1's on the diagonal, and U is an upper triangular matrix. Some authors prefer to introduce the solution of simultaneous equations through the idea of triangular factorisation rather than through the idea of elimination. Others seem to imply that these are really distinct methods, which of course is not true. It is important to realise that the two approaches are mathematically equivalent: we do exactly the same arithmetic whichever approach we use. But triangular factorisation is an important concept in more advanced matrix work, and it is as well to appreciate the dual approach at this stage.

The general 4×4 matrix may be written

$$A = \begin{bmatrix} a_{00} & a_{01} & a_{02} & a_{03} \\ a_{10} & a_{11} & a_{12} & a_{13} \\ a_{20} & a_{21} & a_{22} & a_{23} \\ a_{30} & a_{31} & a_{32} & a_{33} \end{bmatrix},$$

and the corresponding triangular matrices as

$$L = \begin{bmatrix} 1 & 0 & 0 & 0 \\ l_{10} & 1 & 0 & 0 \\ l_{20} & l_{21} & 1 & 0 \\ l_{30} & l_{31} & l_{32} & 1 \end{bmatrix}, \quad U = \begin{bmatrix} u_{00} & u_{01} & u_{02} & u_{03} \\ 0 & u_{11} & u_{12} & u_{13} \\ 0 & 0 & u_{22} & u_{23} \\ 0 & 0 & 0 & u_{33} \end{bmatrix}.$$

On setting $A = LU$ and carrying out the multiplication term by term, we get the sixteen equations below. If we work through these equations in the order given, there is only one unknown element of L or U in each equation, as shown in brackets at the end of the line: thus we can obtain all the elements of L and U in a systematic way.

$$
\begin{array}{ll}
a_{00} = u_{00} & (u_{00}) \\
a_{01} = u_{01} & (u_{01}) \\
a_{02} = u_{02} & (u_{02}) \\
a_{03} = u_{03} & (u_{03}) \\
a_{10} = l_{10} u_{00} & (l_{10}) \\
a_{11} = l_{10} u_{01} + u_{11} & (u_{11}) \\
a_{12} = l_{10} u_{02} + u_{12} & (u_{12}) \\
a_{13} = l_{10} u_{03} + u_{13} & (u_{13}) \\
a_{20} = l_{20} u_{00} & (l_{20}) \\
a_{21} = l_{20} u_{01} + l_{21} u_{11} & (l_{21}) \\
a_{22} = l_{20} u_{02} + l_{21} u_{12} + u_{22} & (u_{22}) \\
a_{23} = l_{20} u_{03} + l_{21} u_{13} + u_{23} & (u_{23}) \\
a_{30} = l_{30} u_{00} & (l_{30}) \\
a_{31} = l_{30} u_{01} + l_{31} u_{11} & (l_{31}) \\
a_{32} = l_{30} u_{02} + l_{31} u_{12} + l_{32} u_{22} & (l_{32}) \\
a_{33} = l_{30} u_{03} + l_{31} u_{13} + l_{32} u_{23} + u_{33} & (u_{33})
\end{array}
$$

For simplicity, we have used only a 4×4 matrix; but you should be able to see how the above equations are extended for a matrix of any size. In fact we obtain the following general formulae for l_{rs} and u_{rs}:

$$
\left.
\begin{array}{ll}
l_{r0} = a_{r0}/u_{00}, & r \geqslant 1, \\
l_{rs} = (a_{rs} - l_{r0} u_{0s} - l_{r1} u_{1s} - \ldots - l_{r,s-1} u_{s-1,s})/u_{ss}, & 1 \leqslant s < r,
\end{array}
\right\} \quad (5.5)
$$

$$
\left.
\begin{array}{ll}
u_{0s} = a_{0s} & \\
u_{rs} = a_{rs} - l_{r0} u_{0s} - l_{r1} u_{1s} - \ldots - l_{r,r-1} u_{r-1,s} & 1 \leqslant r \leqslant s.
\end{array}
\right\} \quad (5.6)
$$

5.2 Computer Implementation

When implementing the triangular factorisation procedure on a computer, the coefficient a_{rs} stored as A(R, S) can be overwritten by l_{rs} for $s < r$ or by u_{rs} for $s \geqslant r$. At the end of the procedure, the array A(R, S) holds the elements of both the matrices L and U, apart from those on the diagonal of L which we know anyway. These matrices are thus available for subsequent use if necessary. Note particularly that we do not waste any of the computer's memory by storing the zero elements of L and U.

The program which follows gives a suitable implementation of the method. The input procedure (lines 2Ø–13Ø), the back substitution (lines 6ØØ–68Ø) and the output procedure (lines 7ØØ–73Ø) are identical with the same lines of the Gaussian elimination program of Section 4.2, and you may find it easier to amend that program rather than enter a completely new program. Lines 4ØØ–47Ø compute l_{rs} using equations (5.5), and lines 5ØØ–56Ø compute u_{rs} using the second equation of (5.6).

```
1Ø REM TRIANGULAR FACTORISATION
2Ø REM ***** INPUT DATA *****
3Ø INPUT"NUMBER OF EQUATIONS";N
4Ø LET N=N-1:PRINT
5Ø DIM A(N,N+1),X(N)
6Ø FOR R=Ø TO N
7Ø FOR S=Ø TO N
8Ø PRINT "A(";R;",";S;") =";
9Ø INPUT A(R,S)
1ØØ NEXT S
11Ø PRINT:PRINT "B(";R;") =";
12Ø INPUT A(R,N+1)
13Ø PRINT:PRINT:NEXT R
2ØØ REM ***** FACTORISATION *****
21Ø FOR R=1 TO N
4ØØ FOR S=Ø TO R-1
41Ø LET P=A(R,S)
42Ø IF S=Ø THEN GOTO 46Ø
43Ø FOR T=Ø TO S-1
44Ø LET P=P-A(R,T)*A(T,S)
45Ø NEXT T
46Ø LET A(R,S)=P/A(S,S)
47Ø NEXT S
5ØØ FOR S=R TO N+1
51Ø LET P=A(R,S)
52Ø FOR T=Ø TO R-1
53Ø LET P=P-A(R,T)*A(T,S)
54Ø NEXT T
55Ø LET A(R,S)=P
56Ø NEXT S
59Ø NEXT R
6ØØ REM ***** BACK SUBSTITUTION *****
61Ø FOR R=N TO Ø STEP -1
62Ø LET P=A(R,N+1)
63Ø IF R=N THEN GOTO 67Ø
64Ø FOR S=R+1 TO N
65Ø LET P=P-A(R,S)*X(S)
66Ø NEXT S
67Ø LET X(R)=P/A(R,R)
68Ø NEXT R
7ØØ REM ***** PRINT RESULTS *****
71Ø FOR R=Ø TO N
72Ø PRINT"X(";R;") =";X(R)
73Ø NEXT R
```

When applied to the set of equations (4.1), this program gives the expected answers:

```
X( Ø ) =-3
X( 1 ) = 1
X( 2 ) = 3
X( 3 ) = 2
```

Since the above program is arithmetically equivalent to the Gaussian elimination program, we must expect it to give the same answers. The time taken for the arithmetical part of the two programs is also identical. However, a significant part of the time taken to solve a set of simultaneous equations is taken up not with arithmetic but with array handling: each time we ask for an element A(R, S) of an array, the computer has to work out where it is stored. In this respect, and in this respect only, the triangular factorisation method is more efficient, and on most computers it turns out to be a little faster than the Gaussian elimination method.

If we try to solve the set of equations (4.6) using the above program, we get the same erroneous answers

```
X( Ø ) =-9.34559666E-Ø3
X( 1 ) = 3.ØØ93458
X( 2 ) = 2.9999999
```

as we did with the Gaussian elimination method. If we wish to use the triangular factorisation method for serious computation we must introduce some sort of pivoting, but this is not so easy as with the elimination method. When solving equations (4.1) by the elimination method, we could, at the second stage, choose our pivot from lines (5), (6) and (7) of Table 4.1. In the factorisation method, lines (6) and (7) would never be obtained explicitly, and our choice of pivots is limited to line (5). This means we can never do total pivoting, and partial pivoting must be based on column interchanges rather than row interchanges. We shall again need to introduce the subsidiary array C(S), as in Section 4.6.

Effectively, in the triangular factorisation method, we ensure that each term on the diagonal of U is the largest term in its row. We should naturally introduce our pivoting between lines 21Ø and 4ØØ of the last program, and at this stage it is in row (R – 1) rather than row (R) that we carry out the pivoting (thus the first time we go through the FOR–NEXT loop commencing in line 21Ø we carry out the pivoting in row (0), although R = 1). This explains the instruction LET Z = R – 1 in line 22Ø below. You will almost certainly find this last point rather difficult to follow, and tricky points such as this can often be the most difficult part of the programming.

The necessary changes to the last program to achieve partial pivoting are as follows:

(i) Change line 1Ø to

```
1Ø REM TRIANGULAR FACTORISATION - PARTIAL PIVOTING
```

(ii) Insert the lines

```
15Ø DIM C(N)
16Ø FOR S=Ø TO N
17Ø LET C(S)=S
18Ø NEXT S
```

(iii) Insert the lines

```
22Ø LET Z=R-1:LET W=Ø
23Ø FOR S=Z TO N
24Ø IF ABS(A(Z,S))>W THEN LET V=S:LET W=ABS(A(Z,S))
25Ø NEXT S
26Ø IF V=Z THEN GOTO 4ØØ
27Ø FOR S=Ø TO N
28Ø LET P=A(S,V)
29Ø LET A(S,V)=A(S,Z)
3ØØ LET A(S,Z)=P
31Ø NEXT S
32Ø LET P=C(V):LET C(V)=C(Z):LET C(Z)=P
```

(iv) Change lines 65Ø and 67Ø to

```
65Ø LET P=P-A(R,S)*X(C(S))
```

and

```
67Ø LET X(C(R))=P/A(R,R)
```

This revised program gives the following near-correct solution to equations (4.6):

```
X( Ø ) = .999999993
X( 1 ) = 2.ØØØØØØØ1
X( 2 ) = 3
```

It also gives satisfactory answers to equations (4.5) and (4.7).

The program just developed is an efficient program for simultaneous equations, which can be recommended for general use. Just occasionally, we may need to use total pivoting, and then we should use the program in Section 4.6.

5.3 The Jacobi Iterative Method

In Chapters 2 and 3 we made effective use of iterative methods to solve equations with only one unknown. It is natural therefore to ask whether they can be used to solve sets of equations with more than one unknown. Consider the set of equations:

$$\left.\begin{aligned} 5x + y + z &= 10 \\ x + 6y - 2z &= 7 \\ x - 3y + 7z &= 16. \end{aligned}\right\} \tag{5.7}$$

The ideas of Chapter 2 suggest the iterative process

$$\left.\begin{aligned} x_{r+1} &= \tfrac{1}{5}(10 - y_r - z_r) \\ y_{r+1} &= \tfrac{1}{6}(7 - x_r + 2z_r) \\ z_{r+1} &= \tfrac{1}{7}(16 - x_r + 3y_r). \end{aligned}\right\} \tag{5.8}$$

A process of this type is known as a **Jacobi iterative method**. Clearly, if it converges it will give the solution of equations (5.7). We implement the method by means of the program below; note that we store x_{r+1}, y_{r+1}, z_{r+1} temporarily as X1, Y1, Z1. The initial values x_0, y_0, z_0 are set as zero in line 2Ø.

```
1Ø REM JACOBI PROCESS
2Ø LET R=Ø:LET X=Ø:LET Y=Ø:LET Z=Ø
3Ø LET X1=(1Ø-Y-Z)/5
4Ø LET Y1=(7-X+2*Z)/6
5Ø LET Z1=(16-X+3*Y)/7
6Ø LET R=R+1:LET X=X1:LET Y=Y1:LET Z=Z1
7Ø PRINT "X(";R;") =";X
8Ø PRINT "Y(";R;") =";Y
9Ø PRINT "Z(";R;") =";Z
1ØØ PRINT:GOTO 3Ø
```

This program gives the output:

```
X( 1 ) = 2
Y( 1 ) = 1.16666667
Z( 1 ) = 2.28571429

X( 2 ) = 1.3Ø952381
Y( 2 ) = 1.595238Ø9
Z( 2 ) = 2.5

....................
....................

X( 28 ) = 1.ØØØØØØØ1
Y( 28 ) = 1.99999999
Z( 28 ) = 2.99999999

X( 29 ) = 1
Y( 29 ) = 2
Z( 29 ) = 3

....................
....................
```

After twenty-nine steps we have obtained what is obviously the correct solution: evidently this is a satisfactory method for solving these equations, though it is doubtful whether it is any quicker than the methods we have used previously. Ideally the program should stop automatically when we have obtained the required accuracy: you may like to make the appropriate changes.

Unfortunately the Jacobi process does not always work! If we apply the method to the set of equations

$$x + 6y - 2z = 7$$

$$x - 3y + 7z = 16$$

$$5x + y + z = 10,$$

which are just equations (5.7) written in a different order, we have to change lines 3Ø–5Ø to

```
3Ø LET X1=7-6*Y+2*Z
4Ø LET Y1=-(16-X-7*Z)/3
5Ø LET Z1=1Ø-5*X-Y
```

We now get the output:

```
....................
....................

X( 54 ) =-6.96679366E+36
Y( 54 ) =-9.53883372E+36
Z( 54 ) = 1.388980Ø36E+37

X( 55 ) = 8.5Ø126Ø94E+37
Y( 55 ) = 3.ØØ872771E+37
Z( 55 ) = 4.437280Ø21E+37

?OVERFLOW ERROR IN 3Ø
```

Clearly, in this case the iterative process has diverged.

Before we can regard the Jacobi process as a serious competitor to the elimination method, we must have some means of determining in advance whether or not the process is going to converge. Consider the set of equations:

$$a_{11}x + a_{12}y + a_{13}z = b_1$$

$$a_{21}x + a_{22}y + a_{23}z = b_2$$

$$a_{31}x + a_{32}y + a_{33}z = b_3.$$

Suppose the true solutions of these equations are x, y, z, and let the iterative values after r steps of the Jacobi process be x_r, y_r, z_r, where

$$x_r = x + \alpha_r, \quad y_r = y + \beta_r, \quad z_r = z + \gamma_r.$$

Further let

$$E_r = \max(|\alpha_r|, |\beta_r|, |\gamma_r|),$$

so that E_r is the largest error after r steps. If E_r gets smaller as r increases, then the process converges; otherwise it diverges. Now in the Jacobi process we set

$$x_{r+1} = (b_1 - a_{12}y_r - a_{13}z_r)/a_{11}$$

and so

$$x + \alpha_{r+1} = (b_1 - a_{12}y - a_{12}\beta_r - a_{13}z - a_{13}\gamma_r)/a_{11}.$$

Since

$$x = (b_1 - a_{12}y - a_{13}z)/a_{11}.$$

we have

$$\alpha_{r+1} = -\frac{a_{12}}{a_{11}}\beta_r - \frac{a_{13}}{a_{11}}\gamma_r.$$

It follows that

$$|\alpha_{r+1}| \leqslant \left|\frac{a_{12}}{a_{11}}\right| |\beta_r| + \left|\frac{a_{13}}{a_{11}}\right| |\gamma_r|$$

$$\leqslant \frac{|a_{12}|}{|a_{11}|} E_r + \frac{|a_{13}|}{|a_{11}|} E_r$$

$$= \frac{|a_{12}| + |a_{13}|}{|a_{11}|} E_r.$$

Similarly we can show that

$$|\beta_{r+1}| \leqslant \frac{|a_{21}| + |a_{23}|}{|a_{22}|} E_r$$

and

$$|\gamma_{r+1}| \leqslant \frac{|a_{31}| + |a_{32}|}{|a_{33}|} E_r.$$

Now if

$$\left.\begin{aligned} |a_{11}| &> |a_{12}| + |a_{13}| \\ |a_{22}| &> |a_{21}| + |a_{23}| \\ |a_{33}| &> |a_{31}| + |a_{32}| \end{aligned}\right\} \qquad (5.9)$$

then

$$|\alpha_{r+1}| < E_r, \quad |\beta_{r+1}| < E_r, \quad |\gamma_{r+1}| < E_r$$

and so

$$E_{r+1} < E_r.$$

Thus we have shown that the process certainly converges if the inequalities (5.9) are satisfied.

These inequalities say in words that for each row of the matrix A, the magnitude of the term on the diagonal exceeds the sum of the magnitudes of the remaining terms. A matrix with this property is said to be **diagonally dominant**. For simplicity, we have considered only a 3×3 matrix in the above work, but the result can be extended to matrices of any size: if the matrix A is diagonally dominant then the Jacobi process for the set of equations $A\boldsymbol{x} = \boldsymbol{b}$ will converge.

We have not proved that the process will not converge if A is not diagonally dominant, and in fact this statement is not true. But it is a very simple matter to check for diagonal dominance and we should not normally embark on a Jacobi process if the matrix were not diagonally dominant; sometimes, of course, we may achieve diagonal dominance by a suitable rearrangement of the equations. Try using the Jacobi process with some simultaneous equations of your own choice. You will find that the method is an attractive alternative to the elimination method when there is substantial diagonal dominance; i.e. when the diagonal terms are very large compared with the remaining terms.

5.4 The Gauss–Seidel Iterative Method

The Jacobi process defined by equations (5.8) is perhaps the most obvious way of using iterative procedures to solve simultaneous equations. But we can also solve equations (5.7) by the iterative process:

$$\begin{aligned} x_{r+1} &= \tfrac{1}{5}(10 - y_r - z_r) \\ y_{r+1} &= \tfrac{1}{6}(7 - x_{r+1} + 2z_r) \\ z_{r+1} &= \tfrac{1}{7}(16 - x_{r+1} + 3y_{r+1}) \end{aligned}$$

In this process we use the *latest available* values of x, y, z instead of the values from the last iterative cycle; thus in the second equation we use the value x_{r+1} just calculated in preference to the previous value x_r. This is known as the **Gauss–Seidel iterative method.**

The Gauss–Seidel process has the important advantage that only one value for each unknown need be stored at any one time, so we can dispense with the temporary variables X1, Y1, Z1 which we needed in the Jacobi process. The following program implements the Gauss–Seidel method for equations (5.7):

```
10 REM GAUSS-SEIDEL PROCESS
20 LET R=0:LET X=0:LET Y=0:LET Z=0
30 LET X=(10-Y-Z)/5
40 LET Y=(7-X+2*Z)/6
50 LET Z=(16-X+3*Y)/7
60 LET R=R+1
70 PRINT "X(";R;") =";X
80 PRINT "Y(";R;") =";Y
90 PRINT "Z(";R;") =";Z
100 PRINT:GOTO 30
```

This program gives the output:

```
X( 1 ) = 2
Y( 1 ) = .833333333
Z( 1 ) = 2.35714286

X( 2 ) = 1.36190476
Y( 2 ) = 1.72539683
Z( 2 ) = 2.83061225

....................
....................

X( 15 ) = 1.00000001
Y( 15 ) = 1.99999999
Z( 15 ) = 3

X( 16 ) = 1
Y( 16 ) = 2
Z( 16 ) = 3

....................
....................
```

It often happens that the Gauss–Seidel method converges more rapidly than the Jacobi process, and of the two the Gauss–Seidel method is usually preferred. It can be shown that the Gauss–Seidel method will converge for the set of equations $\boldsymbol{A}\boldsymbol{x} = \boldsymbol{b}$ if the matrix A is diagonally dominant. We shall not give a proof of this result as it is very similar to the corresponding proof for the Jacobi process, and the interested reader should be able to work out the details.

Exercises

1 Extend your results from Question 2 of Chapter 4 to include the triangular factorisation method, with and without pivoting.

2 Repeat Question 3 of Chapter 4 using the triangular factorisation method with partial pivoting.

3 How can we evaluate $\det(A)$ from the triangular factors of A? Write a program to calculate the determinant of an arbitrary square matrix A (you may assume this determinant is non-zero). Test your program by evaluating

$$\begin{vmatrix} 1 & 2 & 3 & 4 \\ 2 & 1 & 4 & 2 \\ 2 & 5 & 7 & 12 \\ 2 & -2 & 6 & 22 \end{vmatrix}.$$

4 Use the Gauss–Seidel method to solve the set of equations given below, making your program as concise as possible.

$$\begin{aligned} 10u + 3v &= 1 \\ 2u + 11v + 3w &= 2 \\ 2v + 12w + 3x &= 3 \\ 2w + 13x + 3y &= 4 \\ 2x + 14y + 3z &= 5 \\ 2y + 15z &= 6 \end{aligned}$$

5 Write a program to solve the general set of equations $A\boldsymbol{x} = \boldsymbol{b}$ using the Gauss–Seidel method. Your program should first rearrange the equations to make the diagonal terms as large as possible, and stop with a suitable message if diagonal dominance cannot be obtained. Test your program with the set of equations

$$\begin{aligned} w + x - 15y + 4z &= -2 \\ 16w - 2x - 3y + z &= 2 \\ w - x + 3y + 17z &= 9 \\ 2w - 14x + 3y + 2z &= -8. \end{aligned}$$

6

Numerical Integration

6.1 Integration on the Computer

If you have never come across numerical integration before, you may be rather surprised at the idea of integration on a computer. You may envisage that we could somehow write 'x^2' into the computer and get out the answer '$\frac{1}{3}x^3$'. It would certainly be possible to write a program to do that. If the program were sufficiently intricate it could handle integrands such as x^n, $\sin ax$, e^{ax}, as well as deal with simple substitutions and integration by parts. Almost certainly a few programs of this kind have been written. They could be a boon for examination candidates (provided they could smuggle their computer into the examination room), but of little use to the serious mathematician. They could not, for instance, cope with

$$\int e^{-x^2}\,dx$$

since there is no elementary mathematical function whose derivative is e^{-x^2}. Nor could they handle integrands which are calculated only by means of a lengthy computer procedure, or integrands which are obtained from experimental results. These are in fact precisely the types of integration problem for which we use numerical integration.

When we do integration on a computer (or for that matter on a calculator) we restrict ourselves to definite integrals of the form

$$I = \int_a^b f(x)\,dx.$$

Given the values of a and b and the definition of $f(x)$, I is just a number; our aim is to calculate this number to whatever accuracy we require. The value of I is most readily interpreted as an area, and numerical integration is often referred to as **quadrature**, which simply means working out an area. By plotting the graph of $f(x)$ on a piece of graph paper, you could get a rough estimate of I by counting the squares which comprise the appropriate area, making due allowance for part squares. This, in fact, is the most elementary form of numerical integration.

You should be aware from the start that there is an intrinsic difficulty in what we are trying to do. Integration is defined mathematically as a limiting process, and we should need an infinite number of arithmetical operations to work out an integral exactly. On a computer we can perform a large number of operations in a short time, but the number of operations is always finite. At best, therefore, we can hope for an approximation to the value of an integral. We were at pains to point out in Chapter 1 that all numbers in the computer, except simple integers, are merely approximations, but we are now encountering an approximation of a rather different kind. We shall discuss this more fully in Section 6.3.

6.2 The Trapezium Rule

Our first stage in evaluating numerically an integral

$$I = \int_a^b f(x)\,dx$$

is usually to divide the area represented by I into a number of strips, as shown in Fig. 6.1. Usually we take strips of equal width, so that if we take n strips their width is $(b-a)/n = h$, say. For convenience we shall write

$$x_r = a + rh,$$

$$f_r = f(x_r),$$

so that x_r and f_r are as shown in Fig. 6.1.

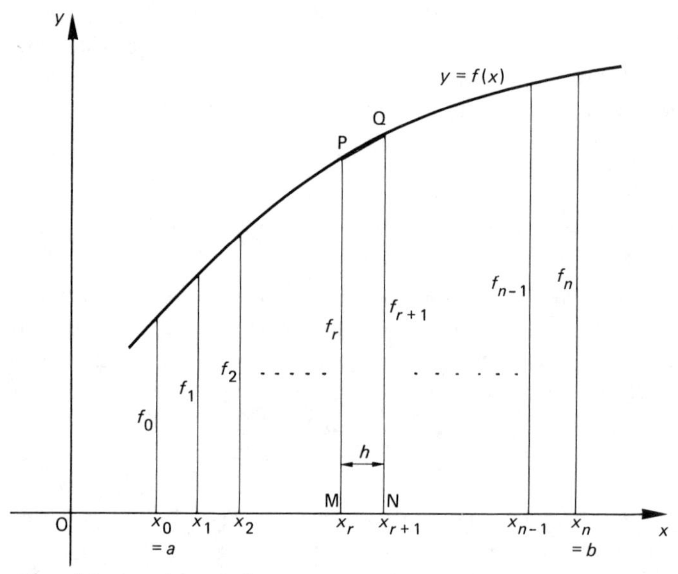

Fig. 6.1 Derivation of the trapezium rule.

We shall look first at the general strip from x_r to x_{r+1}. Provided that h is reasonably small, the area of this strip is approximated by the area of the trapezium NMPQ, in which the arc PQ is replaced by a straight line. (In case you have forgotten your elementary geometry, a trapezium is a quadrilateral with two sides parallel.) The area of a trapezium is the average length of the two parallel sides multiplied by the distance between them; so the area of NMPQ is

$$\tfrac{1}{2}h(f_r + f_{r+1}).$$

The total area under the curve can therefore be approximated by the sum of the areas of n trapezia; that is

$$I \simeq \tfrac{1}{2}h(f_0 + f_1) + \tfrac{1}{2}h(f_1 + f_2) + \tfrac{1}{2}h(f_2 + f_3) + \ldots + \tfrac{1}{2}h(f_{n-2} + f_{n-1}) + \tfrac{1}{2}h(f_{n-1} + f_n)$$

or

$$I \simeq h(\tfrac{1}{2}f_0 + f_1 + f_2 + \ldots + f_{n-1} + \tfrac{1}{2}f_n). \qquad (6.1)$$

This is known as the **trapezium rule.**

The following program uses the trapezium rule to evaluate

$$\int_1^2 e^{-\frac{1}{2}x}\,dx,$$

but it can be used for any other integral by changing lines 2Ø and 3Ø in an obvious way.

```
1Ø REM TRAPEZIUM RULE
2Ø DEF FNF(X)=EXP(-X/2)
3Ø DATA 1,2
4Ø READ A,B
5Ø INPUT"NUMBER OF STRIPS";N
6Ø LET H=(B-A)/N
7Ø LET P=(FNF(A)+FNF(B))/2
8Ø FOR R=1 TO N-1
9Ø LET P=P+FNF(A+R*H)
1ØØ NEXT R
11Ø PRINT "INTEGRAL IS";H*P
```

Typical output from this program is:

```
NUMBER OF STRIPS? 1Ø
INTEGRAL IS .4774Ø1871


NUMBER OF STRIPS? 1ØØ
INTEGRAL IS .47730Ø3431


NUMBER OF STRIPS? 1ØØØ
INTEGRAL IS .4773Ø2448
```

The correct value of the integral is 0.477 302 437, so the last value above is accurate to seven decimal places: but you will find that it takes quite a long time to run. Usually of course we do not know the correct answer. One of the most difficult problems in numerical integration is to decide how large we should take n in order to get the accuracy we require. We shall return to this point later in the chapter.

6.3 The Error in the Trapezium Rule

By using Taylor series we can write

$$f(x_r + t) = a_0 + a_1 t + a_2 t^2 + a_3 t^3 + \ldots,$$

where

$$a_m = \frac{1}{m!} f^{(m)}(x_r) = \frac{1}{m!} f_r^{(m)},$$

and so

$$\int_{x_r}^{x_{r+1}} f(x)\,\mathrm{d}x = \int_0^h f(x_r + t)\,\mathrm{d}t$$

$$= [a_0 t + \tfrac{1}{2}a_1 t^2 + \tfrac{1}{3}a_2 t^3 + \ldots]_0^h$$

$$= a_0 h + \tfrac{1}{2}a_1 h^2 + \tfrac{1}{3}a_2 h^3 + \ldots. \tag{6.2}$$

In the trapezium rule, we approximate this integral by

$$\tfrac{1}{2}h(f_r + f_{r+1}) = \tfrac{1}{2}h\{(a_0) + (a_0 + a_1 h + a_2 h^2 + \ldots)\}$$

$$= a_0 h + \tfrac{1}{2}a_1 h^2 + \tfrac{1}{2}a_2 h^3 + \ldots. \tag{6.3}$$

If we compare equations (6.2) and (6.3), and ignore terms containing h^4 or higher powers of h, we see that the error in our approximation is about

$$-\tfrac{1}{6}a_2 h^3 = -\tfrac{1}{12}h^3 f_r''.$$

(Note the sign: we take the error to be 'what we have to add to our computed answer to get the correct answer'. Some authors use the opposite definition.) This is the error in one strip; the total error in n strips is approximately

$$-\tfrac{1}{12}nh^3 \bar{f}'', \tag{6.4}$$

where $\bar{f}''$ is an 'average' value of f_r''. If we had been more precise with our mathematics, we could show that the error is given exactly by equation (6.4), where $\bar{f}''$ is the value of $f''(x)$ for some value of x between a and b. The values of n and h in equation (6.4) are of course related by $h = (b-a)/n$, so we could write the error either as

$$-\tfrac{1}{12}h^2(b-a)\bar{f}'' \tag{6.5}$$

or as

$$-\tfrac{1}{12}\frac{(b-a)^3}{n^2}\bar{f}''. \tag{6.6}$$

We say that the error is of order h^2 or of order $1/n^2$, abbreviated as $0(h^2)$, $0(1/n^2)$. Thus if we double the number of strips, the error is roughly divided by four.

In the example of the last section, $f(x) = \mathrm{e}^{-\frac{1}{2}x}$, and so $f''(x) = \frac{1}{4}\mathrm{e}^{-\frac{1}{2}x}$. In the interval $(1, 2)$, the value of $f''(x)$ varies between 0.091 97 and 0.151 63. Thus when $n = 10$ the error lies between

$$-\tfrac{1}{12} \times \tfrac{1}{100} \times 0.151\,63 \quad \text{and} \quad -\tfrac{1}{12} \times \tfrac{1}{100} \times 0.091\,97$$

or

$$-0.000\,126\,4 \quad \text{and} \quad -0.000\,076\,6.$$

The actual error is −0.000 099 4, so we have obtained a reasonably good estimate of how accurate the answer is. We know that if we multiply n by 10 we shall divide the error by approximately 100; so we can easily work out how large n has to be to give us any required degree of accuracy. Of course in this problem it is easy to work out

f''; in many cases it is difficult or impossible, and the error formulae (6.5) and (6.6) are then of little immediate use. We shall see in Section 6.6 how we can proceed in these cases.

It is clear from formula (6.6) that the error tends to zero as n tends to infinity. In practice we have to be content with a finite value of n. An error which arises in these circumstances is called a **truncation error**. We shall get a truncation error whenever we try to perform the operations of calculus by numerical means. The truncation error should always be distinguished from the **rounding error**, caused by the computer rounding all numbers (including intermediate results) to a fixed number of binary places. The magnitude of the rounding error is determined by the design of the computer; but we can usually make the truncation error as small as we wish at the expense of increased computer time. There is no point in trying to make the truncation error smaller than the rounding error: the strength of a chain is in its weakest link. But quite often we are satisfied with less accuracy than the computer can provide, and in these circumstances the truncation error might be substantially larger than the rounding error.

6.4 Simpson's Rule

If you have done any numerical integration before, you have almost certainly come across Simpson's rule. Instead of considering the single strip from x_r to x_{r+1}, we consider the two strips from x_{r-1} to x_{r+1} together. As before, we use the Taylor series

$$f(x_r+t)=a_0+a_1t+a_2t^2+a_3t^3+\dots,$$

where

$$a_m=\frac{1}{m!}f_r^{(m)}.$$

Then

$$\int_{x_{r-1}}^{x_{r+1}} f(x)\,\mathrm{d}x=\int_{-h}^{h} f(x_r+t)\,\mathrm{d}t$$

$$=2a_0h+\tfrac{2}{3}a_2h^3+\tfrac{2}{5}a_4h^5+\dots. \tag{6.7}$$

Now

$$f_r=a_0$$

and

$$f_{r-1}+f_{r+1}=2a_0+2a_2h^2+2a_4h^4+\dots.$$

We can easily see that

$$\tfrac{1}{3}h(f_{r-1}+f_{r+1})+\tfrac{4}{3}hf_r=2a_0h+\tfrac{2}{3}a_2h^3+\tfrac{2}{3}a_4h^5+\dots. \tag{6.8}$$

Comparing equations (6.7) and (6.8), we can write

$$\int_{x_{r-1}}^{x_{r+1}} f(x)\,\mathrm{d}x\simeq\tfrac{1}{3}h(f_{r-1}+4f_r+f_{r+1})$$

with error approximately

$$(\tfrac{2}{5}-\tfrac{2}{3})a_4h^5=-\tfrac{4}{15}a_4h^5=-\tfrac{1}{90}h^5f_r^{iv}.$$

If we have divided the interval (a, b) into n strips, where n is even, we can write

$$\int_a^b f(x)\,\mathrm{d}x \simeq \tfrac{1}{3}h\{(f_0+4f_1+f_2)+(f_2+4f_3+f_4)+\ldots+(f_{n-2}+4f_{n-1}+f_n)\}$$

$$=\tfrac{1}{3}h(f_0+4f_1+2f_2+4f_3+2f_4+\ldots+2f_{n-2}+4f_{n-1}+f_n).$$

This is **Simpson's rule**. The error is approximately

$$-\frac{1}{90}\frac{n}{2}h^5\bar{f}^{iv}=-\frac{1}{180}(b-a)h^4\bar{f}^{iv}=-\frac{1}{180}\frac{(b-a)^5}{n^4}\bar{f}^{iv}.$$

The program below uses Simpson's rule to evaluate

$$\int_1^2 \mathrm{e}^{-\frac{1}{2}x}\,\mathrm{d}x,$$

but it can be used to evaluate any other integral by changing lines 2Ø and 3Ø.

```
1Ø REM SIMPSON'S RULE
2Ø DEF FNF(X)=EXP(-X/2)
3Ø DATA 1,2
4Ø READ A,B
5Ø INPUT"NUMBER OF STRIPS";N
6Ø IF N=2*INT(N/2) THEN GOTO 1ØØ
7Ø PRINT"NUMBER MUST BE EVEN"
8Ø GOTO 5Ø
1ØØ LET H=(B-A)/N
11Ø LET P=FNF(A)+FNF(B):LET Z=4
12Ø FOR R=1 TO N-1
13Ø LET P=P+Z*FNF(A+R*H)
14Ø LET Z=6-Z
15Ø NEXT R
16Ø PRINT"INTEGRAL IS";H*P/3
```

Note that we make a check in line 6Ø to ensure that N is even. Note also that the variable Z takes the values 4, 2, 4, 2, ... as we repeat the FOR–NEXT loop; this is achieved very simply by means of line 14Ø. Typical output from the program is:

```
NUMBER OF STRIPS? 4
INTEGRAL IS .4773Ø3Ø83


NUMBER OF STRIPS? 9
NUMBER MUST BE EVEN
NUMBER OF STRIPS? 1Ø
INTEGRAL IS .4773Ø2454


NUMBER OF STRIPS? 2Ø
INTEGRAL IS .4773Ø2438
```

The correct answer is 0.477 302 437. With four strips we have done better than with a hundred strips of the trapezium rule; and with twenty strips we have virtually got machine accuracy. No wonder Simpson's rule is so popular!

For the integral evaluated in the program,

$$f^{iv}(x)=\tfrac{1}{16}\mathrm{e}^{-\frac{1}{2}x},$$

which lies between 0.0230 and 0.0379 in the interval of integration. So with $n = 20$ the error should lie between

$$-\tfrac{1}{180} \times \tfrac{1}{160\,000} \times 0.0379 \quad \text{and} \quad -\tfrac{1}{180} \times \tfrac{1}{160\,000} \times 0.0230;$$

i.e. between $-0.000\,000\,001\,32$ and $-0.000\,000\,000\,80$, which is in accordance with the above results.

6.5 Other Integration Formulae

We shall not prove any further formulae here, but there are many other possibilities. If we take strips three at a time, and let n be a multiple of 3, we get the **three-eighths rule**:

$$\int_a^b f(x)\,dx \simeq \tfrac{3}{8}h(f_0 + 3f_1 + 3f_2 + 2f_3 + 3f_4 + 3f_5 + 2f_6 + \ldots + 2f_{n-3} + 3f_{n-2} + 3f_{n-1} + f_n).$$

This has error

$$-\tfrac{1}{80}(b-a)h^4 \bar{f}^{iv},$$

which is larger than the error in Simpson's rule. For this reason, the three-eighths rule is rarely used.

On the other hand, if we take four strips at a time and let n be a multiple of 4, we get the formula

$$\int_a^b f(x)\,dx \simeq \tfrac{2}{45}h(7f_0 + 32f_1 + 12f_2 + 32f_3 + 14f_4 + 32f_5 + 12f_6 + 32f_7 + 14f_8 + \ldots + 14f_{n-4} + 32f_{n-3} + 12f_{n-2} + 32f_{n-1} + 7f_n), \tag{6.9}$$

which has error

$$-\tfrac{2}{945}(b-a)h^6 \bar{f}^{vi}.$$

Since the error is $O(h^6)$, this formula is more accurate than Simpson's rule, though not quite so convenient to use.

6.6 Repeated Use of Trapezium Rule

In most practical applications of numerical integration, it is not feasible to differentiate the integrand, and the error formulae we have obtained above are of no direct use. We then have to face what is probably the most difficult problem in numerical integration: how do we decide how many strips to use? It is not good enough simply to take a large number of strips and hope for the best: usually this will just waste computer time, but occasionally it will leave us with inadequate accuracy. In the last two sections of this chapter we shall show how to tackle this problem in a systematic and efficient manner.

Suppose we wish to evaluate the integral

$$I = \int_a^b f(x)\,dx,$$

and let I_n be the approximation to I obtained by using the trapezium rule with n strips. Our problem is to decide how large n should be so that I_n approximates I to the required accuracy. One possible approach would be to evaluate in turn $I_1, I_2, I_3, I_4, \ldots$ until two consecutive values agree to the accuracy we need, but this would obviously be very laborious!

A much more sensible approach is to evaluate in turn $I_1, I_2, I_4, I_8, I_{16}, \ldots$. When we do a numerical integration, most of the computer time is taken up with obtaining the function values f_r. The function values needed for I_n are also needed for I_{2n}; so by doubling the number of strips at each stage we do not 'waste' any function values. The procedure actually used is as follows (with $c = b - a$):

(i) Let $J_1 = \frac{1}{2}(f(a) + f(b))$; then $I_1 = cJ_1$.
(ii) Let $J_2 = J_1 + f(a + \frac{1}{2}c)$; then $I_2 = \frac{1}{2}cJ_2$.
(iii) Let $J_4 = J_2 + f(a + \frac{1}{4}c) + f(a + \frac{3}{4}c)$; then $I_4 = \frac{1}{4}cJ_4$.
(iv) Let $J_8 = J_4 + f(a + \frac{1}{8}c) + f(a + \frac{3}{8}c) + f(a + \frac{5}{8}c) + f(a + \frac{7}{8}c)$; then $I_8 = \frac{1}{8}cJ_8$.

and so on. A suitable program is given below:

```
10 REM REPEATED TRAPEZIUM RULE
20 DEF FNF(X)=EXP(-X/2)
30 DATA 1,2
40 READ A,B
50 INPUT"MAX ERROR";Q
60 LET C=(B-A):LET N=1:LET IL=0
70 LET J=(FNF(A)+FNF(B))/2
100 LET I=J*C/N
110 PRINT"I(";N;") =";I
120 IF ABS(I-IL)<Q THEN GOTO 200
130 LET N=2*N:LET IL=I
140 FOR R=1 TO N-1 STEP 2
150 LET J=J+FNF(A+R*C/N)
160 NEXT R
170 GOTO 100
200 PRINT:PRINT"INTEGRAL IS";I
```

Note that J_n and I_n are stored as J and I respectively, the last value of I_n being stored as IL. Typical output from this program is

```
MAX ERROR? .000001
I( 1 ) = .487205051
I( 2 ) = .479785802
I( 4 ) = .477923763
I( 8 ) = .477457799
I( 16 ) = .47734128
I( 32 ) = .477312148
I( 64 ) = .477304865
I( 128 ) = .477303044
I( 256 ) = .477302589

INTEGRAL IS .477302589
```

The systematic approach of the above program has much to recommend it; no decisions are left to the user. On the other hand, because of the relatively large error of the trapezium rule, it could hardly be described as an efficient program. If we ask for a maximum error of 0.000 000 01 the program goes as far as N = 2048, and the computer is kept busy for quite a long time.

6.7 Romberg Integration

The procedure of the last section can be improved enormously by the technique known as **Romberg integration**. We have shown earlier (equation (6.6)) that the error in using the trapezium rule is $0(1/n^2)$. In other words

$$I \simeq I_n + \frac{\alpha}{n^2} \tag{6.10}$$

where α is constant. Similarly

$$I \simeq I_{2n} + \frac{\alpha}{4n^2}. \tag{6.11}$$

By multiplying equation (6.11) by 4 and subtracting equation (6.10), we have

$$3I \simeq 4I_{2n} - I_n$$

or

$$I \simeq I_{2n}^*, \text{ where } I_{2n}^* = \tfrac{1}{3}(4I_{2n} - I_n). \tag{6.12}$$

We can show, with a little algebra, that I_{2n}^* is the approximation to I which we would obtain by using Simpson's rule with $2n$ strips; and we should certainly expect I_{2n}^* to be more accurate than I_{2n}.

We can take this process a stage further. We know that the error in Simpson's rule is $0(1/n^4)$, so for even values of n

$$I \simeq I_n^* + \frac{\beta}{n^4}$$

and

$$I \simeq I_{2n}^* + \frac{\beta}{16n^4}.$$

Thus a better approximation to I is given by I_{2n}^{**}, where

$$I_{2n}^{**} = \tfrac{1}{15}(16I_{2n}^* - I_n^*). \tag{6.13}$$

Proceeding further, we can define

$$I_{2n}^{***} = \tfrac{1}{63}(64I_{2n}^{**} - I_n^{**}),$$
$$I_{2n}^{****} = \tfrac{1}{255}(256I_{2n}^{***} - I_n^{***}),$$

and so on. Hopefully we shall go on getting better and better approximations to I. We may note that I_{2n}^{**} is equivalent to formula (6.9), whilst I_{2n}^{***} and I_{2n}^{****} are equivalent to similar formulae based on eight and sixteen strips.

The technique of Romberg integration is to work out, in sequence, the following approximations to I

$$
\begin{array}{llll}
I_1 & & & \\
I_2 & I_2^* & & \\
I_4 & I_4^* & I_4^{**} & \\
I_8 & I_8^* & I_8^{**} & I_8^{***} \\
\cdots & \cdots & \cdots & \cdots
\end{array}
$$

until two successive values are equal to the accuracy we require. Thus for the integral

$$\int_1^2 e^{-\frac{1}{2}x}\,dx,$$

we get the following table:

0.487 205 051			
0.479 785 802	0.477 312 719		
0.477 923 763	0.477 303 083	0.477 302 441	
0.477 457 799	0.477 302 478	0.477 302 437	0.477 302 437

With only eight strips, we already have nine-figure accuracy; a substantial improvement on the procedure of the last section.

Romberg integration is quite tricky to program. Note first that we need only store one row in the above array at a time. In the program below we store I_n as I(0), I_n^* as I(1), I_n^{**} as I(2), etc., and we also use I and I1 for temporary storage.

```
10 REM ROMBERG INTEGRATION
20 DEF FNF(X)=EXP(-X/2)
30 DATA 1,2
40 READ A,B
50 INPUT"MAX ERROR";Q
60 LET C=(B-A):LET N=1:LET M=0
70 LET J=(FNF(A)+FNF(B))/2
80 DIM I(10):LET I(0)=C*J
100 LET N=2*N:LET M=M+1
110 FOR R=1 TO N-1 STEP 2
120 LET J=J+FNF(A+R*C/N)
130 NEXT R
140 LET I=J*C/N
150 LET S=0:LET T=4
160 LET I1=(I*T-I(S))/(T-1)
170 IF ABS(I-I1)<Q THEN GOTO 250
180 LET I(S)=I:LET I=I1
190 LET S=S+1:LET T=T*4
200 IF S<M THEN GOTO 160
210 LET I(S)=I
220 GOTO 100
250 PRINT"INTEGRAL IS";I1
```

Lines 20–80 initialise the process and set $I(0) = I_1$. Lines 100–220 form the main loop: I_n is evaluated in lines 110–140, and I_n^*, I_n^{**}, . . . in lines 150–200.

Typical output from this program is:

```
MAX ERROR? .ØØØØØØØØ1
INTEGRAL IS .4773Ø2437
```

We have not printed out intermediate results, to avoid making the program too complicated. You might like to extend the program to give some intermediate results, and you might also like to add a few REM statements. You should certainly try running the program with lines 2Ø and 3Ø amended to give some other simple integrals.

The last program can be recommended as a general purpose integral evaluator. We cannot guarantee that it will always work—in calculus it is always possible to think up bizarre functions to shatter any claim of that kind—but it will work in most practical circumstances.

Exercises

1 (a) Evaluate

$$\int_0^1 \sin \tfrac{1}{2}\pi x \, \mathrm{d}x$$

using the trapezium rule with 10, 20, 100, 1000 strips. In each case check that the error accords with that given by formula (6.6).

(b) Estimate the number of strips required to evaluate the above integral to eight decimal places (i) using the trapezium rule, (ii) using Simpson's rule. Carry out the computation in each case, and check that your answer has the required accuracy.

2 (a) Derive the formula

$$\int_{x_{r-2}}^{x_{r+2}} f(x) \, \mathrm{d}x \simeq \tfrac{2}{45}h(7f_{r-2} + 32f_{r-1} + 12f_r + 32f_{r+1} + 7f_{r+2})$$

together with an expression for the error. Hence obtain formula (6.9).

(b) Write a program to evaluate an integral using formula (6.9), and use it to evaluate the integral in Question 1 with various values of n.

3 If we evaluate

$$\int_{a-h}^{a+h} f(x) \, \mathrm{d}x$$

using Simpson's rule with two strips, we have to evaluate $f(x)$ three times and we get an error of order h^5. Show that, by suitable choice of A, B, C and k we can get a formula

$$\int_{a-h}^{a+h} f(x) \, \mathrm{d}x \simeq Af(a - kh) + Bf(a) + Cf(a + kh)$$

which still requires only three function evaluations, but has an error of order h^7. Obtain an approximate expression for this error. Use the formula to evaluate the integral in Question 1.

4 Use Romberg integration to evaluate

$$\text{(a)} \int_1^2 x^{\frac{1}{3}}\,\mathrm{d}x, \qquad \text{(b)} \int_{-1}^2 x^{\frac{1}{3}}\,\mathrm{d}x.$$

Comment on your results. (N.B. the operation X↑Y does not work on most computers if X is negative: you will have to find some way round this.)

7
Differential Equations I

7.1 The General Problem

It is probably true to say that most scientific and technological problems which are solved on a computer are concerned with differential equations in one form or another. The range is vast: from the simple first-order differential equations which we consider in this chapter, to large systems of partial differential equations which tax the capacity of the biggest main-frame machine. The literature on the subject is also vast, and the reader must appreciate that we can only scratch the surface of the topic in an introductory book such as this.

In ordinary mathematics, if we are given a differential equation such as

$$y' = -x^2 y^2,$$

we first find the **general solution**, in this case

$$y = \frac{3}{A + x^3}.$$

The general solution contains an arbitrary constant A. If we are given an **initial condition**, such as $y = 1$ when $x = 0$, we can find the value of A and hence obtain a **particular solution**, namely

$$y = \frac{3}{3 + x^3}.$$

Most differential equations which arise in practice do not admit of a simple analytic solution such as this, and it is for this reason that we need to find numerical solutions. In numerical work we can only hope to obtain a particular solution, and this will consist of a table giving the values of y alongside the corresponding values of x. Thus in the case of a first-order differential equation we need to specify not only the differential equation itself, which may take the general form

$$y' = f(x, y), \tag{7.1}$$

but also an initial condition

$$y = y_0 \quad \text{when} \quad x = x_0,$$

or, more simply, $y(x_0) = y_0$.

Subject to certain mathematical conditions (which need not concern us here since they are nearly always satisfied in practical problems), there is a unique solution of the differential equation (7.1) which passes through (x_0, y_0). We should therefore be able to find, from the information given, the value of y corresponding to any specified value of x. More precisely, if we let $x = x_1 = x_0 + h$, then we should be able to find $y_1 = y(x_1)$.

If we find y_1 by a numerical method there will inevitably be a truncation error. Usually this truncation error depends on h—the larger the value of h, the larger the truncation error—so that in practice we are restricted to small values of h.

However, once we have obtained the point (x_1, y_1) we can start again from the beginning as though this were the initial point, and thus obtain the next point (x_2, y_2). By continuing in this step-by-step manner, we can produce a table of values of x and y over any region we wish. We do not need to use the same interval h in each step, indeed it is often advantageous not to do so.

7.2 Intuitive Approach to Runge–Kutta Methods

The methods which we shall describe here for solving differential equations numerically are known as **Runge–Kutta** methods. Methods of this type have been known for at least a hundred years, but their potential was not fully realised until computers became available.

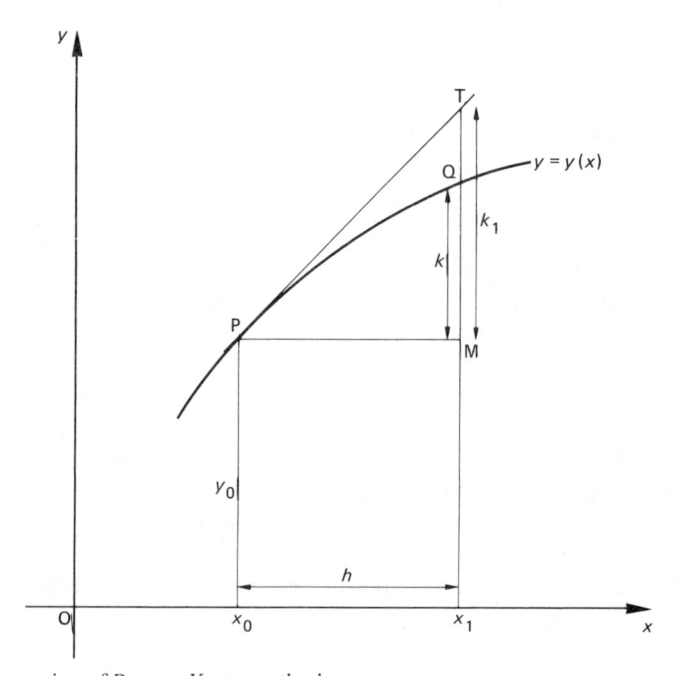

Fig. 7.1 Illustration of Runge–Kutta method

The intuitive basis for a Runge–Kutta method is illustrated in Fig. 7.1. In this diagram the curve labelled $y = y(x)$ is the solution of the differential equation $y' = f(x, y)$ which passes through the point P (x_0, y_0). We want to find the value of $y_1 = y_0 + k$ corresponding to $x = x_1$; in other words we want to find the height MQ. Now although we do not know the position of the curve $y = y(x)$ we do know that at every point on this curve the slope is equal to $f(x, y)$: this is simply the geometric interpretation of the differential equation. Thus the slope of the tangent at P is

$y_0' = f(x_0, y_0)$, and this can be computed since x_0 and y_0 are both known. If h is reasonably small, the tangent PT should not deviate too much from the curve PQ; so the height MT (which by simple geometry is equal to hy_0') should be an approximation to the required height MQ. In other words, a first approximation to k is given by

$$k_1 = hf(x_0, y_0). \tag{7.2}$$

It appears that we might get a better approximation if we used the value of y' at some intermediate point of the curve PQ instead of the value at one end; but since we do not know any of the intermediate points on the curve we cannot do this. We could, however, take

$$k_2 = hf(x_0 + \alpha h, y_0 + \beta k_1), \tag{7.3}$$

where α and β are suitable fractions, as a second approximation to k: but intuition gives us no guidance as to suitable values for α and β. In fact the geometric approach takes us no further, and we have to resort to some rather messy analysis.

7.3 A Second-order Runge–Kutta Method

One of the simplest types of Runge–Kutta method is obtained by defining k_1 and k_2 by equations (7.2) and (7.3), and then approximating k by

$$k \simeq W_1 k_1 + W_2 k_2. \tag{7.4}$$

Our problem is to find suitable values for α, β, W_1, W_2 so that the approximation (7.4) is a reasonably good one.

For simplicity, we use the symbols F, P, Q to denote the values of f, $\partial f/\partial x$, $\partial f/\partial y$ at the point (x_0, y_0). Then

$$k_1 = hF, \tag{7.5}$$

and by using Taylor series in two variables we can write

$$\begin{aligned} k_2 &= h\{F + \alpha hP + \beta k_1 Q + 0(h^2)\} \\ &= hF + \alpha h^2 P + \beta h^2 FQ + 0(h^3). \end{aligned} \tag{7.6}$$

The notation '$+ 0(h^3)$' means that terms containing h^3 and higher powers of h have been neglected.

We also know from Taylor series that

$$y_1 = y(x_0 + h) = y_0 + hy_0' + \tfrac{1}{2}h^2 y_0'' + \ldots,$$

and so

$$k = hy_0' + \tfrac{1}{2}h^2 y_0'' + 0(h^3).$$

Now

$$y' = f(x, y),$$

and so

$$y_0' = F.$$

Further

$$y'' = \frac{d}{dx} f(x, y) = \frac{\partial f}{\partial x}\frac{dx}{dx} + \frac{\partial f}{\partial y}\frac{dy}{dx},$$

so that

$$y_0'' = P + FQ.$$

Thus

$$k = hF + \tfrac{1}{2}h^2(P + FQ) + 0(h^3). \tag{7.7}$$

If you are not too happy about partial differentiation, you will have found the last few steps difficult to follow. However, if you are prepared to accept equations (7.5), (7.6) and (7.7), you should find the rest of this section relatively straightforward.

From the above, our approximation (7.4) is seen to be equivalent to

$$k \simeq (W_1 + W_2)hF + W_2\alpha h^2 P + W_2\beta h^2 FQ + 0(h^3), \tag{7.8}$$

whereas the correct value of k is given by equation (7.7). We want our approximation to be valid for any differential equation, that is for any $f(x, y)$. Clearly the approximation will be valid up to and including terms of order h^2 if

$$\left.\begin{aligned} W_1 + W_2 &= 1, \\ W_2\alpha &= \tfrac{1}{2}, \\ W_2\beta &= \tfrac{1}{2}. \end{aligned}\right\} \tag{7.9}$$

We have four unknown parameters and only three equations to satisfy, so there is some freedom of choice. We can take α to be any number we like (except zero), and then

$$\beta = \alpha, \quad W_1 = 1 - \frac{1}{2\alpha}, \quad W_2 = \frac{1}{2\alpha}.$$

For instance we can take $\alpha = \beta = 1$, $W_1 = W_2 = \tfrac{1}{2}$, and we obtain the following procedure for solving differential equations:

$$\left.\begin{aligned} k_1 &= hf(x_0, y_0) \\ k_2 &= hf(x_0 + h, y_0 + k_1) \\ y_1 &= y_0 + \tfrac{1}{2}(k_1 + k_2) + 0(h^3). \end{aligned}\right\} \tag{7.10}$$

This is a **second-order Runge–Kutta** method: second-order because it is accurate to order h^2, the error being of order h^3. Clearly we could obtain many other second-order methods by choosing different solutions of equations (7.9).

7.4 Computer Implementation of Second-order Method

The following program uses the method (7.10) to solve the differential equation

$$y' = -x^2y^2, \quad y(0) = 1,$$

over the interval $0 \leqslant x \leqslant 2$. It can be adapted to solve any other first-order differential equation by changing lines 81Ø and 91Ø.

```
1Ø REM RUNGE-KUTTA SECOND ORDER METHOD
2Ø REM **** INITIALISE ****
3Ø READ XØ,YØ,XF
4Ø INPUT"INTERVAL";H
1ØØ REM ***** STEP *****
11Ø PRINT XØ,YØ
12Ø IF XØ>XF-.ØØØØØØ1 THEN STOP
13Ø LET X=XØ:LET Y=YØ:GOSUB 91Ø
14Ø LET K1=H*F
15Ø LET X=XØ+H:LET Y=YØ+K1:GOSUB 91Ø
16Ø LET K2=H*F
17Ø LET XØ=X:LET YØ=YØ+(K1+K2)/2
18Ø GOTO 11Ø
8ØØ REM ***** INITIAL X, INITIAL Y, FINAL X *****
81Ø DATA Ø,1,2
9ØØ REM ***** DEFINE F(X,Y) *****
91Ø LET F= -X*X*Y*Y
92Ø RETURN
```

The subroutine in lines 9ØØ–92Ø returns the value $F = f(X, Y)$ for specified values of X and Y. The main computation of each step is carried out in lines 1ØØ–18Ø, and should be easily followed. Line 12Ø tests whether we have reached the final value of x; you should be able to work out for yourself why we use 'X0>XF−.0000001' rather than just 'X0>=XF'!
Typical output from this program is:

```
INTERVAL? .2
 Ø          1
 .2         .996
 .4         .976411613
 .6         .928947137
 .8         .849793601
 1          .746216864
 1.2        .632496125
 1.4        .522443Ø54
 1.6        .42476Ø665
 1.8        .342777322
 2          .276269826
```

and

```
INTERVAL? .Ø1
 Ø          1
 .Ø1        .9999995
 .Ø2        .999997

 .........................
 .........................

 1.97999998               .278755911
 1.98999998               .275727591
 1.99999998               .272734754
```

You will see that the values of x have become slightly inaccurate at the end of this last table: this is the effect of repeated addition of the number 0.01, which the computer cannot store exactly. The correct value of $y(2)$ is $\frac{3}{11} = 0.272\,727\,273$; so we have two decimal places correct using $h = 0.2$, and five using $h = 0.01$. Obviously a major problem is to decide how small to take h in order to get the accuracy we require—a similar problem to the one we faced with numerical integration. We shall return to this point later.

7.5 Higher Order Runge–Kutta Methods

The derivation of higher order Runge–Kutta methods involves some very tedious analysis, and we shall not attempt it here. A simple third-order process is given by

$$\left.\begin{aligned} k_1 &= hf(x_0, y_0) \\ k_2 &= hf(x_0 + \tfrac{1}{2}h, y_0 + \tfrac{1}{2}k_1) \\ k_3 &= hf(x_0 + h, y_0 - k_1 + 2k_2) \end{aligned}\right\} \quad (7.11)$$

$$y_1 = y_0 + \tfrac{1}{6}(k_1 + 4k_2 + k_3) + 0(h^4)$$

and a simple fourth-order process by

$$\left.\begin{aligned} k_1 &= hf(x_0, y_0) \\ k_2 &= hf(x_0 + \tfrac{1}{2}h, y_0 + \tfrac{1}{2}k_1) \\ k_3 &= hf(x_0 + \tfrac{1}{2}h, y_0 + \tfrac{1}{2}k_2) \\ k_4 &= hf(x_0 + h, y_0 + k_3) \\ y_1 &= y_0 + \tfrac{1}{6}(k_1 + 2k_2 + 2k_3 + k_4) + 0(h^5). \end{aligned}\right\} \quad (7.12)$$

In both cases the processes are not unique, and many other processes are possible.

You will have noticed that the second-order process uses two ks, the third-order process uses three, and the fourth-order process uses four. It looks as though a pattern is emerging, but in fact this is not so: a fifth-order process requires at least six ks, and a sixth-order process at least eight. These methods are extremely complicated, and for this reason we are usually content to use a fourth-order method. In fact the fourth-order process (7.12) is the one most frequently used.

It is easy to implement process (7.12) by amending the program of the last section. Line 1Ø is changed to

```
1Ø REM RUNGE-KUTTA FOURTH ORDER METHOD
```

and lines 15Ø–22Ø become

```
15Ø LET X=XØ+H/2:LET Y=YØ+K1/2:GOSUB 91Ø
16Ø LET K2=H*F
17Ø LET Y=YØ+K2/2:GOSUB 91Ø
18Ø LET K3=H*F
19Ø LET X=XØ+H:LET Y=YØ+K3:GOSUB 91Ø
2ØØ LET K4=H*F
21Ø LET XØ=X:LET YØ=YØ+(K1+2*(K2+K3)+K4)/6
22Ø GOTO 11Ø
```

This amended program gives the output

```
INTERVAL? .2
 Ø             1
 .2            .997339983
 .4            .979111274
 .6            .932833319
 .8            .8542Ø8317
 1             .7499915Ø7
 1.2           .634512221
 1.4           .522287751
 1.6           .422787425
 1.8           .339694949
 2             .27275Ø784
```

and

```
INTERVAL? .Ø1
 Ø             1
 .Ø1           .999999667
 .Ø2           .999997333

 .........................
 .........................

 1.97999998              .278748445
 1.98999998              .27572Ø117
 1.99999998              .272727276
```

Using this process we get four decimal places with $h = 0.2$, and eight with $h = 0.01$; a substantial improvement on the second-order method.

7.6 Estimation of the Truncation Error

When we are using the fourth-order Runge–Kutta method, we know that the truncation error is of order h^5. It is possible to obtain a precise expression for this error, as we did in numerical integration. But this expression is a very complicated one, involving high order partial derivatives, and has very little practical use. So we have to find some other method of estimating the truncation error.

Suppose we start at (x_0, y_0) and proceed to (x_1, y_1) and then to (x_2, y_2) using the same interval h in each case. We obtain not the correct value of y_1 but $y_1 - E$, where E is the truncation error. In the second step we start with the value $y_1 - E$ instead of y_1; if we assume that the truncation error in this step is still approximately E, then we shall obtain roughly $y_2 - 2E$ instead of y_2. Now suppose that we go directly from (x_0, y_0) to (x_2, y_2) using an interval $2h$. Since the truncation error is of order h^5, its value should now be approximately $32E$; so we shall obtain roughly $y_2 - 32E$ instead of y_2. This is illustrated in the table below:

x	Correct y	Computed y interval h	Computed y interval $2h$
x_0	y_0	y_0	y_0
x_1	y_1	$y_1 - E$	↓
x_2	y_2	$y_2 - 2E$	$y_2 - 32E$

The value of E is roughly one-thirtieth of the difference between the two computed values of y_2; or the error in the first value of y_2 is roughly one-fifteenth of this difference.

Thus in solving the differential equation

$$y' = -x^2 y^2, \quad y(0) = 1$$

with interval 0.2 we obtain the results

$$y(0.2) = 0.997\ 339\ 983,$$

$$y(0.4) = 0.979\ 111\ 274.$$

Repeating with interval 0.4, we obtain

$$y(0.4) = 0.979\ 084\ 908.$$

The error in the first value of $y(0.4)$ is therefore approximately

$$\tfrac{1}{15}(0.979\ 111\ 274 - 0.979\ 084\ 908) = 0.000\ 001\ 758.$$

The correct value of $y(0.4)$ is 0.979 112 271, so the actual error is 0.000 000 997. We certainly cannot claim that our error estimate is an accurate one! It does, however, give us a realistic estimate of how many decimal places are correct, and in practice this is all we should demand of an error estimate.

It is quite possible to write a program which automatically adjusts the interval h so as to keep the truncation error (or at least our estimate of it) within prescribed bounds. After every two steps we estimate the truncation error as explained above. If the error is too large, we reduce h appropriately and repeat the two steps. If the error is too small we need not repeat the steps, but we start the next two steps with a larger value of h. We shall give a program which does precisely this in the next chapter.

It is important to realise that the error we have just been discussing is the truncation error in a particular step. We usually refer to it as the **local truncation error** to remind us that it is 'local' to this step. A complete analysis of the errors in the numerical solution of a differential equation involves very much more than this. In the second step of a solution we are solving the differential equation with initial condition $(x_1, y_1 - E)$, whereas we really require the solution with initial condition (x_1, y_1); the two solutions may diverge markedly from one another. The problem of estimating **global** errors—the differences between the true solution and the computed solution after many steps—is much less tractable than that of estimating local errors. Suffice it to say here that keeping the local error within bounds, although very desirable on its own account, gives no guarantee of global accuracy.

Exercises

1 If your computer has high resolution graphics, modify the program of Section 7.5 so that instead of printing out a table of x and y, it plots a graph of y against x on the screen. The range of values of x from X0 to XF should cover the whole width of the screen; you will need to specify the vertical scale for each problem. Test your program with some simple differential equations, for example

(i) $y' = -xy, \quad y(0) = 1, \quad 0 \leqslant x \leqslant 3;$
(ii) $y' = -x/y, \quad y(-4) = 3, \quad -4 \leqslant x \leqslant 4.$

2 Find a formula for estimating the local truncation error in the second-order process of Section 7.3. Write a program to implement this process which prints out an estimate of the truncation error after every two steps.

3 Use the program of Section 7.5 to obtain the solution of the differential equation

$$y' = y - 2e^{-x}, \quad y(0) = 1$$

over the region $0 \leqslant x \leqslant 10$. Compare your answers with the theoretical solution $y = e^{-x}$, and explain any discrepancies.

8

Differential Equations II

8.1 Systems of Equations

So far we have considered only single differential equations, but practical problems are often concerned with systems of simultaneous differential equations. The simplest system of this kind is the pair of equations

$$\left.\begin{aligned} y' &= f(x, y, z), \\ z' &= g(x, y, z). \end{aligned}\right\} \qquad (8.1)$$

In this system x is the independent variable, and we require to find y and z as functions of x, subject to initial conditions of the form

$$y(x_0) = y_0, \quad z(x_0) = z_0.$$

There is no real difficulty in applying Runge–Kutta methods to problems of this type. The fourth-order method of the last chapter becomes

$$\begin{aligned}
k_1 &= hf(x_0, y_0, z_0) \\
l_1 &= hg(x_0, y_0, z_0) \\
k_2 &= hf(x_0 + \tfrac{1}{2}h, y_0 + \tfrac{1}{2}k_1, z_0 + \tfrac{1}{2}l_1) \\
l_2 &= hg(x_0 + \tfrac{1}{2}h, y_0 + \tfrac{1}{2}k_1, z_0 + \tfrac{1}{2}l_1) \\
k_3 &= hf(x_0 + \tfrac{1}{2}h, y_0 + \tfrac{1}{2}k_2, z_0 + \tfrac{1}{2}l_2) \\
l_3 &= hg(x_0 + \tfrac{1}{2}h, y_0 + \tfrac{1}{2}k_2, z_0 + \tfrac{1}{2}l_2) \\
k_4 &= hf(x_0 + h, y_0 + k_3, z_0 + l_3) \\
l_4 &= hg(x_0 + h, y_0 + k_3, z_0 + l_3)
\end{aligned}$$

$$\begin{aligned}
y_1 &= y_0 + \tfrac{1}{6}(k_1 + 2k_2 + 2k_3 + k_4) + 0(h^5) \\
z_1 &= z_0 + \tfrac{1}{6}(l_1 + 2l_2 + 2l_3 + l_4) + 0(h^5).
\end{aligned}$$

Although the above process looks rather fearsome at first sight, its structure should be reasonably easy to follow; indeed it is not too difficult to see how the procedure can be extended to a system of any number of equations. But the notation is obviously going to be very cumbersome for a large system of equations, and it is useful to tidy this up a little before proceeding.

First we rewrite the system (8.1) in the extended form

$$\left.\begin{aligned} x' &= 1, \\ y' &= f(x, y, z), \\ z' &= g(x, y, z). \end{aligned}\right\} \tag{8.2}$$

The first equation merely states the obvious fact that $\mathrm{d}x/\mathrm{d}x = 1$; but by including this equation in the system we put all three variables on an equal footing, and remove the special status of the variable x. We next introduce the vector $\boldsymbol{y}$ defined by

$$\boldsymbol{y} = \begin{bmatrix} x \\ y \\ z \end{bmatrix}$$

(be careful to distinguish between $\boldsymbol{y}$ and y) and the vector function $\boldsymbol{f}$ defined by

$$\boldsymbol{f}(\boldsymbol{y}) = \begin{bmatrix} 1 \\ f(x, y, z) \\ g(x, y, z) \end{bmatrix}.$$

The system (8.2) can then be written in the simple vector form

$$\boldsymbol{y}' = \boldsymbol{f}(\boldsymbol{y}). \tag{8.3}$$

If we now define the vector $\boldsymbol{k}_r$ by

$$\boldsymbol{k}_r = \begin{bmatrix} h \\ k_r \\ l_r \end{bmatrix},$$

the fourth-order Runge–Kutta method can be written in the simple form

$$\begin{aligned} \boldsymbol{k}_1 &= h\boldsymbol{f}(\boldsymbol{y}_0) \\ \boldsymbol{k}_2 &= h\boldsymbol{f}(\boldsymbol{y}_0 + \tfrac{1}{2}\boldsymbol{k}_1) \\ \boldsymbol{k}_3 &= h\boldsymbol{f}(\boldsymbol{y}_0 + \tfrac{1}{2}\boldsymbol{k}_2) \\ \boldsymbol{k}_4 &= h\boldsymbol{f}(\boldsymbol{y}_0 + \boldsymbol{k}_3) \end{aligned}$$

$$\boldsymbol{y}_1 = \boldsymbol{y}_0 + \tfrac{1}{6}(\boldsymbol{k}_1 + 2\boldsymbol{k}_2 + 2\boldsymbol{k}_3 + \boldsymbol{k}_4).$$

In order to extend the method to a system of different size, we merely have to change the dimensions of the various vectors; in other words the above process can be used to solve a system of first-order differential equations of any size. When we implement the process on a computer, these vectors are simply stored as arrays of appropriate dimensions.

8.2 Computer Implementation

We give below a program to solve any system of first-order differential equations:

```
10 REM RUNGE-KUTTA METHOD FOR SYSTEM OF EQUATIONS
20 REM ***** INITIALISE *****
30 READ N
40 DIM Y(N),Y0(N),F(N),K1(N),K2(N),K3(N)
50 FOR R=0 TO N:READ Y0(R):NEXT R
60 READ XF
70 INPUT"INTERVAL";H
100 REM ***** STEP *****
110 FOR R=0 TO N:PRINT Y0(R):NEXT R:PRINT
120 IF Y0(0)>XF-.0000001 THEN STOP
130 FOR R=0 TO N:LET Y(R)=Y0(R):NEXT R
140 GOSUB 910
150 FOR R=0 TO N:LET K1(R)=H*F(R)
160 LET Y(R)=Y0(R)+K1(R)/2:NEXT R
170 GOSUB 910
180 FOR R=0 TO N:LET K2(R)=H*F(R)
190 LET Y(R)=Y0(R)+K2(R)/2:NEXT R
200 GOSUB 910
210 FOR R=0 TO N:LET K3(R)=H*F(R)
220 LET Y(R)=Y0(R)+K3(R):NEXT R
230 GOSUB 910
235 FOR R=0 TO N
240 LET Y0(R)=Y0(R)+(K1(R)+2*(K2(R)+K3(R))+H*F(R))/6
245 NEXT R
250 GOTO 110
790 REM ***** DATA *****
800 DATA 2:REM NUMBER OF DEPENDANT VARIABLES
810 DATA 0,0,1:REM INITIAL VALUES
820 DATA 1:REM FINAL X
900 REM ***** DEFINE F *****
910 LET F(0)=1
920 LET F(1)=-2*Y(0)*Y(1)+Y(2)
930 LET F(2)=-2*Y(0)*Y(2)-Y(1)
940 RETURN
```

As written the program solves the pair of equations

$$\left.\begin{aligned} y' &= -2xy + z, \\ z' &= -2xz - y, \end{aligned}\right\} \tag{8.4}$$

subject to the initial conditions $y(0) = 0$, $z(0) = 1$, over the interval $0 \leqslant x \leqslant 1$. For any other system of equations we have to change lines 800 onwards. Lines 800–820 give the necessary data, which should be self-explanatory. Lines 900 onwards give a subroutine for evaluating the vector function $\boldsymbol{f}$: Y(0), Y(1), Y(2) denote the values of x, y, z and F(0), F(1), F(2) the expressions for x', y', z'. The rest of the program should be reasonably easy to follow. Note that we do not use an array K4(); it is not needed, and it is always as well to avoid filling the store with unnecessary arrays.

Typical output from this program is

```
INTERVAL? .2
 Ø
 Ø
 1

 .2
 .19Ø878933
 .941642667

 .4
 .33184955
 .784882983

 .6
 .393954563
 .5758Ø7475

 .8
 .3782540Ø93
 .367333132

 1
 .3Ø95Ø6Ø57
 .19873Ø16
```

The theoretical solution of the pair of equations (8.4) is

$$y = e^{-x^2} \sin x, \quad z = e^{-x^2} \cos x,$$

which gives $y(1) = 0.309\ 559\ 876$, $z(1) = 0.198\ 766\ 110$; thus we have about four decimal places correct. You may like to run the program with some other intervals: you should get full machine accuracy with $h = 0.01$.

8.3 Higher Order Equations

So far we have considered only first-order differential equations, but of course we often want to solve equations of higher order. Consider, for instance, the second-order equation

$$\left.\begin{aligned} &y'' + 3y' + 2y = 2e^{-3x} \\ &y(0) = 1, \quad y'(0) = -2. \end{aligned}\right\} \tag{8.5}$$

If we denote y' by z, the differential equation can be written

$$z' + 3z + 2y = 2e^{-3x}.$$

We can thus regard equation (8.5) as the system of equations

$$\left.\begin{aligned} &x' = 1 \\ &y' = z \\ &z' = -2y - 3z + 2e^{-3x}, \end{aligned}\right\} \tag{8.6}$$

with initial conditions $y(0) = 1$, $z(0) = -2$. We can solve this system using the program of the last section. To obtain a solution in the region $0 \leqslant x \leqslant 1$, we simply change lines

```
81Ø DATA Ø,1,-2:REM INITIAL VALUES
```

and

```
92Ø LET F(1)=Y(2)
93Ø LET F(2)=-2*Y(1)-3*Y(2)+2*EXP(-3*Y(Ø))
```

Typical output from the program is now

```
INTERVAL? .2
 Ø
 1
-2

 .2
 .697458546
-1.125Ø1219

 .4
 .522462149
-.6758172Ø1

 .6
 .413159388
-.442824697

 .8
 .338339756
-.318Ø85146

 1
 .282469166
-.246859951
```

The theoretical solution of equation (8.5) is

$$y = e^{-x} - e^{-2x} + e^{-3x},$$

giving $y(1) = 0.282\,331\,226$; so in this problem an interval $h = 0.2$ gives us rather less than four decimal accuracy.

We are now in a position to solve a substantial proportion of differential equations which arise in practice. There are, however, some problems which the above program cannot handle. In most textbooks on differential equations, you will come across problems such as

$$\left.\begin{aligned} y'' + 3y' + 2y &= 2e^{-3x} \\ y(0) = 1, \quad y(1) &= 0. \end{aligned}\right\} \qquad (8.7)$$

This differential equation can be rewritten as the system (8.6), but we cannot use the above program because we do not know the value of $z(0)$. Problems such as equation (8.7) are known as **boundary value problems**, as distinct from the **initial value problems** which we have considered here. Generally speaking, boundary value problems are more difficult to solve than initial value problems, and they are beyond the scope of this book. One obvious approach is to guess the value of $z(0)$ and proceed by trial-and-error until we get a solution with the required value of $y(1)$. It is possible to develop this approach into a sound numerical method—see Exercise 3 at the end of this chapter.

8.4 Interval Adjustment

The next program, which again solves a system of equations, is designed to keep automatic control over the local truncation error. More precisely, the program chooses an interval h so that the estimated local truncation error is less than a specified tolerance T. A pair of steps is carried out using interval h and repeated using interval $2h$; an error estimate can then be obtained by the method described in the last chapter. The error estimate is now obtained as the difference between two vectors, and so is a vector itself; the largest element (in magnitude) of this vector is denoted by E. If $E < T$ the two steps are accepted. If not, h is repeatedly divided by 2, and E by 32, until E is less than T; the two steps are then recomputed using this smaller value of h. If $E < T/40$ then h is repeatedly multiplied by 2, and E by 32, until $E \geqslant T/40$; this larger value of h is then used as the initial interval for the next two steps.

In the program, lines 600–730 contain a subroutine which carries out a single step of the fourth-order Runge–Kutta method using an interval HA. The initial values for this step are specified in the array YA( ), and the final values are returned in the same array. Lines 30–80 carry out the necessary initialisation; line 110 prints out the vector Y0( ) on entry to a pair of steps; and line 120 tests whether we have reached the finishing value XF. Line 130 is included to ensure that we stop precisely at XF and do not overshoot. In lines 140–210 we carry out two steps with interval H, storing the results in the arrays Y1( ) and Y2( ), and a single step with interval 2H. Lines 310–340 compute E, the largest element of the error estimate vector, and lines 400–530 carry out the necessary interval adjustment. Lines 800 onwards specify the problem being solved, as in the last program.

```
10 REM RUNGE-KUTTA METHOD WITH INTERVAL ADJUSTMENT
20 REM ***** INITIALISE *****
30 READ N
40 DIM Y(N),Y0(N),Y1(N),Y2(N),YA(N),F(N),K1(N),K2(N),K3(N)
50 FOR R=0 TO N:READ Y0(R):NEXT R
60 READ XF
70 LET H=1
80 INPUT"TOLERANCE";T
100 REM ***** NEXT TWO STEPS *****
110 FOR R=0 TO N: PRINT Y0(R):NEXT R:PRINT
120 IF Y0(0)>XF-.0000001 THEN STOP
130 IF Y0(0)+2*H>XF-.0000001 THEN LET H=(XF-Y0(0))/2
140 PRINT "H =";H:LET HA=H
150 FOR R=0 TO N:LET YA(R)=Y0(R):NEXT R
160 GOSUB 610
170 FOR R=0 TO N: LET Y1(R)=YA(R):NEXT R
180 GOSUB 610
190 LET HA=2*H
200 FOR R=0 TO N:LET Y2(R)=YA(R):LET YA(R)=Y0(R):NEXT R
210 GOSUB 610
300 REM ***** COMPUTE ERROR ESTIMATE *****
310 LET E=0
320 FOR R=0 TO N: LET D=ABS(YA(R)-Y2(R))/15
330 IF D>E THEN LET E=D
340 NEXT R
350 IF E<T THEN GOTO 510
400 REM ***** REDUCE H *****
410 LET H=H/2:LET E=E/32
420 IF E>=T THEN GOTO 410
430 GOTO 140
```

```
5ØØ REM ***** ACCEPT STEPS *****
51Ø FOR R=Ø TO N:PRINT Y1(R):LET YØ(R)=Y2(R):NEXT R:PRINT
52Ø IF E>T/4Ø THEN GOTO 11Ø
53Ø LET H=2*H:LET E=32*E:GOTO 52Ø
6ØØ REM ***** SINGLE STEP *****
61Ø FOR R=Ø TO N:LET Y(R)=YA(R):NEXT R
62Ø GOSUB 91Ø
63Ø FOR R=Ø TO N:LET K1(R)=HA*F(R)
64Ø LET Y(R)=YA(R)+K1(R)/2:NEXT R
65Ø GOSUB 91Ø
66Ø FOR R=Ø TO N:LET K2(R)=HA*F(R)
67Ø LET Y(R)=YA(R)+K2(R)/2:NEXT R
68Ø GOSUB 91Ø
69Ø FOR R=Ø TO N:LET K3(R)=HA*F(R)
7ØØ LET Y(R)=YA(R)+K3(R):NEXT R
71Ø GOSUB 91Ø
715 FOR R=Ø TO N
72Ø LET YA(R)=YA(R)+(K1(R)+2*(K2(R)+K3(R))+HA*F(R))/6
725 NEXT R
73Ø RETURN
79Ø REM ***** DATA *****
8ØØ DATA 2:REM NUMBER OF DEPENDANT VARIABLES
81Ø DATA Ø,Ø,1:REM INITIAL VALUES
82Ø DATA 1:REM FINAL X
9ØØ REM ***** DEFINE F *****
91Ø LET F(Ø)=1
92Ø LET F(1)=-2*Y(Ø)*Y(1)+Y(2)
93Ø LET F(2)=-2*Y(Ø)*Y(2)-Y(1)
94Ø RETURN
```

As written the program solves the pair of equations (8.4), and the table below shows the values obtained for $y(1)$ and $z(1)$ using various tolerance levels:

Tolerance	$y(1)$	$z(1)$	No. of steps
0.01	0.308 767 531	0.195 144 653	2
0.000 1	0.309 551 694	0.198 762 310	8
0.000 001	0.309 559 379	0.198 765 947	16
Theoretical	0.309 559 876	0.198 766 110	—

In each case we obtain an accuracy appropriate to the specified tolerance: but it must be remembered that we are controlling only the local errors, and that we can give no guarantee of the global accuracy in our final results. In the above trials the program uses the same interval h throughout the region $0 \leqslant x \leqslant 1$. If the tabulation is continued over a larger region—say by changing the number 1 in line 82Ø to 5—we find that the interval is increased when we get beyond about $x = 3$. By this stage the graphs of y and z have flattened out, and are very close to zero.

We would not claim that the program just given is the last word in programming Runge–Kutta methods: there are a number of refinements which might help to avoid difficulties when dealing with 'awkward' problems. Nevertheless a program such as this does enable a casual computer user to solve a differential equation numerically without worrying too much about what happens inside the computer. You should try using the program on as many differential equations as possible (you will find plenty listed as exercises in mathematics textbooks!) Occasionally you will find a problem for which the program fails, and you should ask yourself why it has failed: the most likely cause is instability, which we discuss in the remainder of this chapter.

8.5 Stability and Instability

The question of global accuracy is one of the most difficult aspects of the numerical solution of differential equations, and well beyond the scope of the present volume. However, there are some circumstances in which global errors can get completely out of control, behaviour which is known as **instability**. The consequences of instability are potentially serious, and anyone who attempts to solve a differential equation numerically should have some awareness of the dangers.

For simplicity, we shall consider the simple differential equation

$$y' = \lambda y. \tag{8.8}$$

The general solution of this equation is

$$y = A\mathrm{e}^{\lambda x};$$

if $\lambda > 0$ the solution increases rapidly as x increases, whereas if $\lambda < 0$ the solution tends to zero as x increases. In practical problems we are not often interested in explosive situations; more often we are interested in a physical system which is tending to a steady state, or at least moving within fixed bounds. Thus if equation (8.8) arises in practice, it is likely that λ will be negative: and the least we should hope is that our numerical solution tends to zero as x increases.

Let us now see what happens when we apply the Runge–Kutta fourth-order method to equation (8.8). Writing $w = h\lambda$ we get

$$k_1 = h\lambda y_0 = wy_0,$$

$$k_2 = h\lambda(y_0 + \tfrac{1}{2}wy_0) = y_0(w + \tfrac{1}{2}w^2),$$

$$k_3 = h\lambda\{y_0 + \tfrac{1}{2}y_0(w + \tfrac{1}{2}w^2)\} = y_0(w + \tfrac{1}{2}w^2 + \tfrac{1}{4}w^3),$$

$$k_4 = h\lambda\{y_0 + y_0(w + \tfrac{1}{2}w^2 + \tfrac{1}{4}w^3)\} = y_0(w + w^2 + \tfrac{1}{2}w^3 + \tfrac{1}{4}w^4).$$

So we find

$$y_1 = y_0 + \tfrac{1}{6}\{wy_0 + 2y_0(w + \tfrac{1}{2}w^2) + 2y_0(w + \tfrac{1}{2}w^2 + \tfrac{1}{4}w^3) + y_0(w + w^2 + \tfrac{1}{2}w^3 + \tfrac{1}{4}w^4)\}$$

or

$$y_1 = y_0(1 + w + \tfrac{1}{2}w^2 + \tfrac{1}{6}w^3 + \tfrac{1}{24}w^4). \tag{8.9}$$

It is not difficult to see that theoretically we should have

$$y_1 = y_0\,\mathrm{e}^{w},$$

and, provided that w is reasonably small, it looks as though equation (8.9) should give a good approximation to the theoretical solution.

Since $w = h\lambda$ and h is essentially positive, we are interested primarily in the case where w is negative. The graph of

$$\phi(w) = 1 + w + \tfrac{1}{2}w^2 + \tfrac{1}{6}w^3 + \tfrac{1}{24}w^4$$

for negative values of w is shown in Fig. 8.1. We can see from the graph that $|\phi(w)| < 1$ only if w lies between 0 and -2.8 approximately (more precisely, the lower limit is $-2.785\ 293\ 56$). If w lies between these limits, then by virtue of equation (8.9) y_1 will be smaller than y_0, and our solution will tend to zero as x increases. On the other hand, if w lies outside these limits, y_1 will be larger than y_0, and the numerical solution will increase without limit as x increases. There is no problem if $w > 0$, for in this case we expect an increasing solution. But if $w < -2.8$ we get an *increasing* numerical solution, although the theoretical solution is *decreasing*. This is what we mean by numerical instability: clearly in these circumstances the numerical solution is of no use whatsoever.

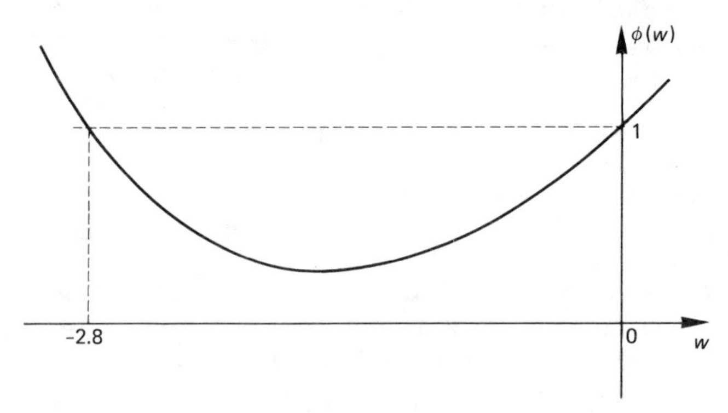

Fig. 8.1 Graph of $\phi(w) = 1 + w + \frac{1}{2}w^2 + \frac{1}{6}w^3 + \frac{1}{24}w^4$

The message, then, is this: when solving the differential equation (8.8) with a negative value of λ, we must choose h so that $h\lambda > -2.8$; otherwise we are asking for trouble. Of course we do not usually want to solve an equation as simple as (8.8) numerically, and the stability situation for the more general equation

$$y' = f(x, y)$$

is not nearly so easy to determine. But generally speaking we shall get a decreasing solution only if $h(\partial f/\partial y)$ lies between 0 and -2.8.

8.6 The Problem of Stiffness

First we will use the program of Section 8.4 to solve the simple differential equation

$$\left.\begin{aligned} y' &= -y \\ y(0) &= 1, \end{aligned}\right\} \qquad (8.9)$$

over the region $0 \leqslant x \leqslant 10$. To do this, we merely change lines 8ØØ onwards in the program to read

```
8ØØ DATA 1:REM NUMBER OF DEPENDANT VARIABLES
81Ø DATA Ø,1:REM INITIAL VALUES
82Ø DATA 1Ø:REM FINAL X
9ØØ REM DEFINE F
91Ø LET F(Ø)=1
92Ø LET F(1)=-Y(1)
93Ø RETURN
```

By virtue of the work of the last section, we must use an interval h less than 2.8. If we run the program with a relatively large tolerance, say $T = 0.01$, the program initially chooses an interval 0.5, which it subsequently increases to 1 and then to 2. Clearly we are staying comfortably within the stability region. In practice, we rarely get stability problems when using a Runge–Kutta method to solve a single first-order differential equation.

The real problem arises with higher order or simultaneous equations. If, for instance, we have a second-order differential equation whose general solution is

$$y = A\,e^{\lambda x} + B\,e^{\mu x}$$

then for stability we have to ensure that both $h\lambda$ and $h\mu$ lie between -2.8 and 0. There can be significant difficulties if λ and μ are of appreciably different size. We shall illustrate this by solving the differential equation

$$\left.\begin{aligned} y'' + 1001y' + 1000y = 0 \\ y(0) = 1, \quad y'(0) = -1 \end{aligned}\right\} \qquad (8.10)$$

over the region $0 \leqslant x \leqslant 10$. To do this we amend lines 8ØØ onwards in the program of Section 8.4 to read

```
8ØØ DATA 2:REM NUMBER OF DEPENDANT VARIABLES
81Ø DATA Ø,1,-1:REM INITIAL VALUES
82Ø DATA 1Ø:REM FINAL X
9ØØ REM DEFINE F
91Ø LET F(Ø)=1
92Ø LET F(1)=Y(2)
93Ø LET F(2)=-1ØØ1*Y(2)-1ØØØ*Y(1)
94Ø RETURN
```

The correct solution of equation (8.10) is $y = e^{-x}$, the same as the solution of equation (8.9). In this case, however, the solution is obtained much more slowly. If we use a tolerance $T = 0.01$, there is some initial fluctuation in the interval length, but eventually the program settles on $h = 0.001\,953\,125$, with occasional steps using $h = 0.003\,906\,25$. With such small step-lengths, it takes a very long time to complete the computation.

The general solution of equation (8.10) is

$$y = A\,e^{-x} + B\,e^{-1000x},$$

and for a stable solution we need

$$-2.8 < -h < 0$$

and

$$-2.8 < -1000h < 0;$$

these inequalities are both satisfied only if $h < 0.0028$. This explains why the program uses the small interval 0.001 953 125. We could get an accurate tabulation of $y = e^{-x}$ with a much larger interval, but the small interval is necessary to ensure stability.

Equations such as (8.10) are known as **stiff differential equations**. The only objection to using Runge–Kutta methods for stiff problems is that they are inordinately slow. Equation (8.10) is only moderately stiff compared with many modern technological problems, and some fast and efficient methods for handling highly stiff problems have been developed in recent years. Ideally we should like a program which automatically chooses an appropriate method for a stiff or non-stiff problem, and automatically controls the global errors. This is the direction of much present day research.

Exercises

1 (a) Change the program of Section 8.2 so as to solve the differential equation

$$y'' + (20 + \tanh x)y' + (20 + \operatorname{sech} x)y = 0$$
$$y(0) = 0, \quad y'(0) = 1,$$

over the region $0 \leqslant x \leqslant 4$. Run the program with various values of h, and show that the solution becomes unstable if h is greater than about 0.14. Find a value of h which gives full machine accuracy, and tabulate the solution to this accuracy. (To avoid a lengthy print-out, modify the program so that the solution is printed only for $x = 0(0.5)4$.)

(b) Tabulate the solution of the differential equation

$$y'' + (20 + \tanh x)y' + (20 + \operatorname{sech} x)y = x^2$$
$$y(0) = 1, \quad y'(0) = 0,$$

to machine accuracy for $x = 0(0.5)4$.

(c) By taking a linear combination of the solutions obtained in (a) and (b) above, find the solution of the boundary value problem

$$y'' + (20 + \tanh x)y' + (20 + \operatorname{sech} x)y = x^2$$
$$y(0) = 1, \quad y(4) = \tfrac{1}{2},$$

Note: this approach can be used for any *linear* boundary value problem.

2 Use the program of Section 8.4 to solve the differential equation

$$y'' - 5(1 - y^2)y' + y = 0$$
$$y(0) = 2, \quad y'(0) = 0,$$

over the region $0 \leqslant x \leqslant 10$. You will find that the program makes substantial changes in the interval as the computation proceeds. If you draw a graph of y against x (or better, get the computer to plot it on the screen) you should be able to see the reasons for this behaviour.

3 (a) Consider the differential equation

$$y'' = 6y^2$$
$$y(1) = 1, \quad y'(1) = u.$$

The value of y when $x = 2$ obviously depends on u: denote this value by $\psi(u)$. Write a subroutine, based on the program of Section 8.4, to compute the value of $\psi(u)$ to about four decimal places for any specified value of u.

(b) Starting with the values $u_0 = 0$, $u_1 = -1$, use the secant method to find the value of u for which $\psi(u) = \frac{1}{4}$. Hence tabulate the solution of the boundary value problem

$$y'' = 6y^2$$
$$y(1) = 1, \quad y(2) = \tfrac{1}{4},$$

to about four decimal places for $1 \leqslant x \leqslant 2$.

Note: this method, known as the **shooting method**, is a standard procedure for solving boundary value problems.

Answers to Exercises

Chapter 1

1 23 455.999 7, 0.000 298 431 109.
If you get any other answer for the second root, you are probably subtracting two nearly equal numbers.

3 A suitable program would be:

```
10 REM EXPONENTIAL SERIES
20 INPUT"X";X
30 LET Q=1E-9
40 LET R=0:LET T=1:LET S=1
50 LET R=R+1:LET T=T*X/R:LET S=S+T
60 IF ABS(T)>Q THEN GOTO 50
70 PRINT "SUM IS";S
```

In this program T denotes the current term and S the sum so far. The program works well for positive and small negative values of x. For large negative values, the series contains large terms of oscillating sign which nearly cancel out to give a small answer; this answer will be highly inaccurate.

4 (i) 0.000 000 333 333 333, (ii) 0.000 000 044 456 985 3.

5 (i) $\{0.031\,62, 0.054\,78\}$, (ii) $\{0.099\,96, 0.172\,99\}$,
(iii) $\{3.597, 4.147\}$, (iv) $\{63.3, 172.5\}$,
(v) $\{-0.002\,52, -0.001\,78\}$.

6 (a) $x = \{1.4293, 1.4395\}$, $y = \{2.0025, 2.0382\}$
$x = \{31.3, 458.0\}$ $y = \{-1600.0, -106.7\}$.

7 £10 760, with a possible error of about $\pm$£150.

Chapter 2

1 (i) Converges for all x_0 to 0.739 085 133.
(ii) Converges to 0.618 033 989 if $-1.618\,033\,989 < x_0 < 1$, to 1 if $x_0 = 1$; diverges otherwise.
(iii) Diverges for all x_0.
(iv) Converges for all x_0 to $-1.324\,717\,96$. (You will get an error message if you take $x_0 > 1$, because your computer does not like working out fractional powers of negative numbers: you should be able to overcome this difficulty.)

2 The process converges to $x = 5$, $y = -7$, the solution of the pair of simultaneous equations $3x + 2y = 1$, $x + 3y = -16$.

3 The root is $-1.371\,134\,33$.

Chapter 3

1 The following subroutine returns the value C, where A$(C) = B$, or gives an error message if B$ is not found:

```
1ØØØ REM BINARY SEARCH
1Ø1Ø LET A=Ø: LET B=N+1
1Ø2Ø LET C=INT((A+B)/2)
1Ø3Ø IF B$=A$(C) THEN RETURN
1Ø4Ø IF B$<A$(C) THEN LET B=C:GOTO 1Ø6Ø
1Ø5Ø LET A=C
1Ø6Ø IF B-A>1 THEN GOTO 1Ø2Ø
1Ø7Ø PRINT B$;" NOT FOUND"
1Ø8Ø RETURN
```

2 The roots are −2.702 061 37, 0.342 185 053, 2.210 083 94.

3 (i) 1.432 032 24,
(ii) 4.669 584 78 (but most of the starting values suggested lead to a different root),
(iii) 7.828 439 32.

4 A suitable program is

```
1Ø REM FALSE POSITION
2Ø DEF FNF(X)=EXP(-X)-X
3Ø INPUT"INITIAL LIMITS";A,B
4Ø LET F=FNF(A):LET G=FNF(B):LET Q=1E-9
5Ø IF F*G<Ø THEN GOTO 8Ø
6Ø PRINT"F(A) AND F(B) HAVE SAME SIGN:TRY AGAIN"
7Ø GOTO 3Ø
8Ø LET D=C
9Ø LET C=B-(A-B)*G/(F-G)
1ØØ LET H=FNF(C)
11Ø IF ABS(C-D)<Q THEN GOTO 16Ø
12Ø IF F*H<Ø THEN LET B=C:LET G=H:GOTO 14Ø
13Ø LET A=C:LET F=H
14Ø PRINT"ROOT BETWEEN";A;"AND";B
15Ø GOTO 8Ø
16Ø PRINT"ROOT IS";C
```

The last suggested pair of starting values gives very slow convergence.

5 Theoretically the roots are 2, 3.5, 3.5; the program gives 2, 3.499 921 71, 3.500 078 29.

Chapter 4

1 Replace lines 6Ø–13Ø by

```
6Ø FOR R=Ø TO N
7Ø LET P=Ø
8Ø FOR S=Ø TO N
9Ø LET A(R,S)=RND(1)
1ØØ LET P=P+(S+1)*A(R,S)
11Ø NEXT S
12Ø LET A(R,N+1)=P
13Ø NEXT R
```

3 You will need to change the dimensions of the arrays to A(N, N + M), X(N, M). You will also need to change the input/output routines, the back substitution routine, and the upper limit of the FOR statement in lines 27Ø and 42Ø. You should get the answer

$$\begin{bmatrix} 4 & 1 & 2 \\ 3 & 1 & 0 \\ 2 & 1 & 0 \\ 1 & 1 & 1 \end{bmatrix}.$$

4
$$\begin{bmatrix} 183 & -31 & -65 & 10 \\ 40 & -7 & -14 & 2 \\ -110 & 19 & 39 & -6 \\ 17 & -3 & -6 & 1 \end{bmatrix}$$

Chapter 5

3 The determinant is the product $u_{00}u_{11}\ldots u_{nn}$. Your program should carry out the triangular factorisation (omitting the right-hand sides) and then evaluate the required product. The given determinant has value −2. If you use partial pivoting, remember that a column interchange changes the sign of the determinant.

4 A suitable program, which automatically stops when nine decimal accuracy is obtained, is:

```
1Ø REM GAUSS-SEIDEL
2Ø DIM X(7):LET Q=1E-9
3Ø LET E=Ø
4Ø FOR R=1 TO 6
5Ø LET P=(R-2*X(R-1)-3*X(R+1))/(R+9)
6Ø PRINT "X(";R;") =";P
7Ø LET W=ABS(P-X(R)):LET X(R)=P
8Ø IF W>E THEN LET E=W
9Ø NEXT R:PRINT
1ØØ IF E>Q THEN GOTO 3Ø
```

The solution is:

```
X( 1 ) = .Ø63Ø8789Ø5
X( 2 ) = .123Ø4Ø365
X( 3 ) = .17346ØØ67
X( 4 ) = .224132823
X( 5 ) = .2464510Ø56
X( 6 ) = .367139859
```

5 $\frac{1}{4}, \frac{3}{4}, \frac{1}{3}, \frac{1}{2}$.

Chapter 6

1 (b) (i) 4530, (ii) 43.

3 $A = C = \frac{5}{9}h,\ B = \frac{8}{9}h,\ k = \sqrt{\frac{3}{5}}$.

The error is approximately $\frac{1}{15\,750}h^7 f^{vi}$.

4 Both integrals have the value 1.139 881 57, but you will have difficulty with (b). In (b), the integrand has an infinite derivative within the region of integration; we cannot therefore use Taylor series within this region, and this invalidates the derivation of the integration formula.

Chapter 7

2 The error estimate is the same as for the fourth-order process, except that the fraction one-fifteenth is replaced by one-third.

3 The general solution of this differential equation is

$$y = A\,e^{x} + e^{-x}.$$

For the given initial condition we should have $A = 0$, but truncation and rounding errors lead to a non-zero value of A after the first step. The rapid increase in e^{x} soon makes this the dominant term.

Chapter 8

1 The required numerical values are:

x	(a)	(b)	(c)
0.0	0	1	1
0.5	0.032 372 437	0.614 870 90	0.630 908 77
1.0	0.019 009 230	0.371 776 58	0.381 194 08
1.5	0.011 307 820	0.250 990 91	0.256 593 00
2.0	0.006 781 799	0.209 227 50	0.212 587 33
2.5	0.004 086 620	0.223 485 68	0.225 510 26
3.0	0.002 469 200	0.281 020 50	0.282 243 79
3.5	0.001 494 256	0.374 489 55	0.375 229 83
4.0	0.000 905 086	0.499 551 60	0.500 000 00

Note that (c) = (b) + 0.495 417 50 × (a).

2 You will see that y starts to decrease sharply near $x = 5$; then the graph turns a sharp corner and y starts to increase again. In this region a small interval is needed, but elsewhere the interval can be relatively large. The correct value of $y(10)$ is −1.158 701 6.

3 (b) The correct solution of the boundary value problem is $y = x^{-2}$, corresponding to $u = -2$.

Bibliography

The following books are recommended for further reading:

(1) R. L. Burden, J. D. Faires and A. C. Reynolds, *Numerical Analysis*, Prindle, Weber and Schmidt, Boston, Mass., Second edition, 1981.
(2) A. M. Cohen et al., *Numerical Analysis*, McGraw-Hill, London, 1973.
(3) S. D. Conte and C. de Boor, *Elementary Numerical Analysis, an algorithmic approach*, McGraw-Hill Kogakusha, Tokyo, Second edition, 1972.
(4) W. S. Dorn and D. D. McCracken, *Introductory Finite Mathematics with Computing*, John Wiley and Sons, New York, 1976.
(5) C. F. Gerald, *Applied Numerical Analysis*, Addison-Wesley, Reading, Mass., Second edition, 1978.
(6) F. B. Hildebrand, *Introduction to Numerical Analysis*, Tata McGraw-Hill, New Delhi, Second edition, 1974.
(7) L. W. Johnson and R. D. Riess, *Numerical Analysis*, Addison-Wesley, Reading, Mass., 1977.
(8) *Modern Computing Methods*, HMSO: NPL Notes on Applied Science, No. 16, 1961.

Index